Dust Explosions in the
Process Industries

Dust Explosions in the Process Industries

Editor

Sanjay Khanwelkar

Dust Explosions in the Process Industries
Edited by **Sanjay Khanwelkar**

Printed in 2017

ISBN: 978-1-68117-357-3

Library of Congress Control Number: 2015941549

© 2016 by
SCITUS Academics LLC,
616, Corporate Way, Suite 2, 4766,
Valley Cottage, NY 10989

www.scitusacademics.com

This book contains information obtained from highly regarded resources. Copyright for individual articles remains with the authors as indicated. All chapters are distributed under the terms of the Creative Commons Attribution License, which permits unrestricted use, distribution, and reproduction in any medium, provided the original author and source are credited.

Notice

Reasonable efforts have been made to publish reliable data and views articulated in the chapters are those of the individual contributors, and not necessarily those of the editors or publishers. Editors or publishers are not responsible for the accuracy of the information in the published chapters or consequences of their use. The publisher believes no responsibility for any damage or grievance to the persons or property arising out of the use of any materials, instructions, methods or thoughts in the book. The editors and the publisher have attempted to trace the copyright holders of all material reproduced in this publication and apologize to copyright holders if permission has not been obtained. If any copyright holder has not been acknowledged, please write to us so we may rectify.

Contents

Preface vii

Chapter 1   **Dust Explosion Prevention and Mitigation, Status and Developments in Basic Knowledge and in Practical Application** 1

Rolf K. Eckhoff

Chapter 2   **Experimental Study of Constant Volume Sulfur Dust Explosions** 39

Joseph Kalman, Nick G. Glumac, and Herman Krier

Chapter 3   **Industry Specific Dust Explosion Likelihood Assessment Model with Case Studies** 71

Junaid Hassan, Faisal Khan, Paul Amyotte, and Refaul Ferdous

Chapter 4   **Experiment-based Investigations of Magnesium Dust Explosion Characteristics** 107

Niansheng Kuai, Jianming Li, Zhi Chen, Weixing Huang, Jingjie Yuan, and Wenqing Xu

Chapter 5   **Modelling the Effect of Particle Size on Dust Explosions** 143

A. Di Benedetto, P. Russo, P. Amyotte, and N. Marchand

Chapter 6   **A Reaction Engineering Approach to Modeling Dust Explosions** 167

Vimlesh Kumar Bind, Shantanu Roy, and Chitra Rajagopal

Chapter 7   **Correlations for Flame Speed and Explosion Overpressure of Dust Clouds inside Industrial Enclosures** 201

M. Silvestrini, B. Genova, and F.J. Leon Trujillo

Chapter 8	Dust Explosions: CFD Modeling as a Tool to Characterize the Relevant Parameters of the Dust Dispersion 255
	Carlos Murillo, Olivier Dufaud, Nathalie Bardin-Monnier, Omar López, Felipe Munoz, Laurent Perrin

Citations 295

Index 299

Preface

Dust explosions are common and costly in a wide array of industries such as petrochemical, food, paper and pharmaceutical. It is imperative that practical and theoretical knowledge of the origin, development, prevention and mitigation of dust explosions is imparted to the responsible safety manager. The material in this book offers an up to date evaluation of prevalent activities, testing methods, design measures and safe operating techniques. Also provided is a detailed and comprehensive critique of all the significant phases relating to the hazard and control of a dust explosion.

Editor

Chapter 1

Dust Explosion Prevention and Mitigation, Status and Developments in Basic Knowledge and in Practical Application

Rolf K. Eckhoff[1,2]

[1]Department of Physics and Technology, University of Bergen, Allegaten 55, N-5007 Bergen, Norway

[2]Tyréns AB, 205 19 Malmö, Sweden

ABSTRACT

Right from the early days of the process industries, continuous efforts have been made to develop and improve measures for prevention and mitigation of dust explosions in these industries. Nevertheless this hazard continues to threaten industries that manufacture, use and/or handle powders and dusts of a wide range of combustible materials. To improve methods for predicting explosion development in real industrial plant has been one major challenge. Hence, during the last years comprehensive numerical simulation codes, for addressing this problem, have been developed. Progress has also been made in other areas, for example, ignition source prevention. The importance of adopting inherently safer process design, by building on firm knowledge in powder science and technology, and of systematic education/training of personnel, is also emphasized.

INTRODUCTION

Table 1 gives an overview of the most important methods currently used for preventing and mitigating dust explosions in the process industries. In dust explosion prevention and mitigation, as in many other challenges encountered by the process industries, there is an inevitable conflict between the short-term needs of the users of knowledge and technology and the long-term strive by researchers for the "perfect" solution. Industry will always need practicable tools and means that can be implemented more or less immediately. On the other hand, however, industrial pragmatism must not block the constant strive for better solutions based on improved basic understanding of the phenomena involved. A main aim of the present paper is to elucidate how fundamental research can promote further development of the practical means for preventing and mitigating dust explosions in industry that are listed in Table 1.

Table 1: Schematic overview of means of preventing and mitigating dust explosions

Explosion prevention		Explosion mitigation
Preventing explosive dust clouds	Preventing ignition sources	
Inerting of dust clouds by N_2, CO_2 and rare gases	Smouldering combustion in dust, dust fires	Explosion-pressure resistant construction

Intrinsic inerting of dust cloud by combustion gases	Other types of open flames (e.g., hot work)	Explosion isolation (sectioning)
Inerting of dust cloud by adding inert dust	Hot surfaces (electrically or mechanically heated)	Explosion venting
Keeping dust conc. outside explosive range	Heat from mechanical impact (metal sparks and hot-spots)	Automatic explosion suppression
Inherently safer process design	Electric sparks and arcs and electrostatic discharges	Partial inerting of dust cloud by inert gas
		Good housekeeping (dust removal/cleaning)

IN-DEPTH KNOWLEDGE—A POWERFUL AND ESSENTIAL TOOL IN ASSESSING AND CONTROLLING DUST EXPLOSION HAZARDS IN PRACTICE

Over the last 20 years, there has been a gradual shift in approach in dust explosion prevention and mitigation, from simple schematic design methods, toward more sophisticated ones opening up for increased flexibility and tailoring. During the same time period the appreciation of the benefits to be harvested from cross-fertilization between systematic research and practical applications has been growing. Advanced numerical models are starting to play an increasingly important role in solving practical design problems. The development of such models requires detailed experimental and theoretical studies of the relevant physical and chemical aspects. Table 2 summarizes some fundamental research topics that are essential for further development of the preventive and mitigatory methods and design tools, that are indicated in Table 1. For example, basic understanding of flame propagation processes in dust clouds is the key to adequate design of practical mitigatory measures such as systems for dust explosion venting, suppression, and isolation. In Section 3 some work on basic aspects of dust flames will be reviewed.

Table 2: Fundamental aspects addressed in dust explosion research

Dust cloud formation processes	Dust cloud ignition processes	Flame propagation processes in dust clouds	Blast waves generated by burning dust clouds
Inter-particle forces in dust deposits (cohesion)	General theories for ignition of single particles and clouds	Single-particle ignition and combustion in hot oxidizer gas	Blast wave properties as a function of properties of burning dust clouds
Entrainment of particles from dust deposits by shock waves passing across the deposit surface	Ignition by smouldering combustion in dust layers/deposits	Laminar and turbulent flames in dust clouds	Effects of blast waves on humans and mechanical structures
	Ignition by hot surfaces		
Entrainment of particles from dust deposits by turbulent gas flows Transport of dust particles in turbulent gas flows	Ignition by flying burning metal particles	Mechanisms of heat transfer (conduction, convection, radiation)	Ability of blast waves from dust explosions to transform dust layers into explosive dust clouds (coupled to first column of table)
	Ignition by electric sparks and arcs		
	Ignition by electrostatic discharges	Limit conditions for flame propagation in dust clouds (particle properties, dust conc., oxygen conc., geometry).	
	Ignition by hot gas jets		
	Ignition by shock waves		

Measurement and characterization of turbulence in dust clouds	Ignition by hot-spots from focused light beams	Acceleration of flames in dust clouds by turbulence mechanisms	
Measurement and characterization of spatial distribution of particles in dust clouds	Influences on dust cloud ignition sensitivity of cloud properties (composition, size, shape of particles, dust concentration, composition, turbulence, temperature and pressure of gas phase)	Detonation phenomena in dust clouds	

FLAME PROPAGATION IN DUST CLOUDS

Differences between Premixed Gases and Dust Clouds

In dust clouds, as opposed to premixed gases, inertial forces can produce fuel concentration gradients (displacement of particles in relation to gas phase). Furthermore, thermal radiation may contribute significantly to the heat transfer from the flame to the unburnt cloud, depending on the type of particle material (e.g., light metals). More work is needed to explore the role of thermal radiation in the development and course of dust explosions. Some papers discussing various central issues are those by Lee et al. [1] and Wolanski [2]. Much research work has been done on various aspects of combustion of liquid sprays and mists [3], which is in part also relevant even in the context of dust explosions.

Ignition and Combustion of Single Particles

Through the years a substantial amount of work has been conducted on various aspects of the ignition and combustion of single particles. A comprehensive review is given by Eckhoff [4]. Two more recent examples will be given here. One is the general study by Frolov et al. [5] on the effect of transient heat transfer on ignition of solid particles; the other is the investigation by Fedorov and Shulgin [6] on the stability of the process of ignition of small magnesium particles.

Flames in Dust Clouds

It has often been assumed that the laminar burning velocity of a given dust cloud is a basic combustion property of the cloud, which is closely related also to the burning velocities at various defined levels of turbulence and hence to the flame propagation through that type of cloud at large. An excellent recent contribution to improve understanding of the nature of laminar dust flames was given by Dahoe et al. [7].

Adequate submodels of flame propagation in turbulent dust clouds are essential in comprehensive numerical codes for dust explosion propagation.

In the case of gases, extensive experimental research programmes have been conducted to resolve basic flame acceleration mechanisms in obstructed geometries. Central contributors are Moen et al. [8], Hjertager et al. [9], and Bakke and van Wingerden [10]. The fundamental studies of Rzal-Rebière and Veyssière [11] provide significant insight in possible differences between turbulent combustion of premixed gases and dust clouds. They investigated the interaction of a laminar maize starch/air flame with an obstacle, namely, a sphere, a disk, or an annulus.

One very interesting possibility would be to perform dust explosion experiments in large-scale experimental facilities used in previous extensive gas explosion experiments, for example, in the experiments by Moen et al. [8] on the influence of turbulence-generating baffles. By repeating these experiments with explosive dust clouds, and comparing the results with those found previously for gas, both important similarities and important discrepancies between turbulent dust and gas explosions could be disclosed.

Significant differences between combustion of premixed gases and dust clouds also exist on the microscopic scale. For example, the basic microscopic turbulence mechanisms that promote the combustion process must be identified. The results of Mitgau [12] and Mitgau et al. [13] indicate that more efficient replacement of gaseous reaction products by fresh air round each particle can be a strong basic turbulent combustion enhancement mechanism.

Cashdollar and Zlochower [14] measured flame temperatures and maximum explosions pressures in dust explosions with powders of a wide range of metals and sulphur, boron, and carbon. These data are useful in developing numerical models of dust flames.

Comprehensive Mathematical Models for Turbulent Flame Propagation in Dust Clouds

Kjäldman [15] was one of the pioneers in applying computational fluid dynamics (CFD) to turbulent dust explosion propagation. Subsequent contributions were made by Rose et al. [16], Smirnov et al. [17], Bielert and Sichel [18], Wörsdörfer et al. [19], Korobeinikov et al. [20], Zhong et al. [21], and Kosinski et al. [22]. Di Benedetto and Russo [23] presented a thermokinetic model of dust explosion propagation for natural and synthetic organic dusts. The model was based on the assumption that the devolatilization/pyrolysis step is very fast, and that the subsequent gas phase explosion is controlling the explosion rate. In developing a comprehensive numerical code for dust explosion simulation, corresponding existing codes for gas explosion simulation constitute a logical starting point. The comprehensive FLame ACceleration Simulator (FLACS) code, originally developed by Hjertager et al. [9] is currently being used as a basis for developing the corresponding dust explosion code

Dust Explosion Simulation Code (DESC). van Wingerden et al. [24], Arntzen et al. [25], Hansen et al. [26], and Siwek et al. [27] presented dust explosion simulations using preliminary versions of the DESC code. Skjold et al. [28–33] and Skjold [34], using improved versions of the same code, presented results from extensive simulations of dust explosion experiments performed in full scale process equipment, including a silo of 236m^3. Most probably this type of comprehensive numerical computer simulation code will become a future tool for predicting the course of complex dust explosion scenarios encountered in the process industries, for example, explosions propagating through a series of consecutive process units connected with ducts.

GENERATION OF EXPLOSIVE DUST CLOUDS IN PROCESS PLANT AND MEANS FOR THEIR PREVENTION

A Historical Perspective

Nearly 130 years ago Professor Weber, one of the pioneers of dust explosion research, stressed the importance of accounting for dust cohesion and dust dispersibility when considering the possibility of generation of explosive dust clouds. In his excellent paper on the ignitability and explosibility of wheat flour, Weber [35] emphasized that the cohesion of the flour, which is caused by interparticle adhesion, has a strong influence on the ability of the flour to disperse into explosive dust clouds. Weber suggested that two large dust explosion disasters, one in Szczecin (Stettin) and one in München, were mainly due to the high dispersibility of the flour. He also demonstrated, using simple but convincing laboratory experiments, that the dispersibility, or dustability, of wheat flour increased as its moisture content decreased. A global definition of dust dispersibility is given in Eckhoff [4, Chapter 3].

Generation of Primary Dust Clouds inside Process Equipment

In order for an explosive dust cloud to be formed from a layer/deposit, the layer/deposit must be exposed to a process that suspends the particles in the air to the extent that the dust concentration drops into the explosive range. Most often such dispersion of dust to form explosive clouds takes place intentionally inside process equipment, for example, by handling and transportation in various process equipment (e.g., mills, dryers, mixers, bucket elevators and other conveyors, silos, filters, cyclones, and connecting ducts). It is foreseen that in a not-too-distant future comprehensive numerical codes will be available for predicting the dust cloud structures (spatial distributions of effective particle size, dust concentration, turbulence, and global flow) that will be generated in various practical scenarios in industry. Knowing this initial cloud structure is essential both for predicting the ignition sensitivity of the cloud with regard to various ignition sources and for predicting the course of development of the primary explosion that will result from ignition. Therefore, adequate information about initial dust cloud structures is essential for realistic assessment of the dust explosion risk in a process plant. However, the development of adequate numerical models of dust cloud structures is not far beyond its infancy, and information of practical use is scarce. The works of Hauert et al. [36] and Kosinski et al. [37] constitute valuable initial contributions. More recent contributions are by Kosinski and Hoffmann [38], Kosinski et al. [39], and Ilea et al. [40]. However, the problem addressed is very complex and more experimental, and theoretical work is needed.

Generation of Secondary Dust Clouds inside and/or Outside Process Equipment by Blast Waves from Primary Dust Explosions

The blast wave from a primary dust explosion can generate secondary explosive clouds ahead of the flame by entraining dust deposits and layers there. Lebecki et al. [41], being primarily concerned with coal mine explosions, investigated this process experimentally in a 100 m long gallery of cross-section $3m^2$. Kauffman et al. [42] and Austin et al. [43] summarized their extensive research on blast-wave entrainment of dust layers in long tubes, whereas Boiko and Poplavski [44] studied the effect of the dust concentration in a dust cloud behind a shock wave, on the acceleration of the cloud. Data from this kind of work are essential in the development of comprehensive dust explosion codes. Klemens et al. [45] presented a mathematical model for simulating the process of entrainment of dust particles from a dust layer, by the gas flow behind a shock or a rarefaction wave passing across the layer. The above mentioned subsequent contributions by Kosinski and Hoffmann [38], Kosinski et al. [39], and Ilea et al. [40] are indeed relevant also in the present context. The mathematical models of Fedorov and Gosteev [46] and Fedorov and Fedorova [47] describing the initial stage of the entrainment of single dust particles from a dust layer by a gas flow passing across the layer are also important contributions.

Dust Dispersibility Tests

Various test methods have been proposed for evaluating the ease with which dust clouds can be produced from deposits and layers of powders/dusts [4, Chapter 7], [48–51].

Inherently Safer Process Design to Prevent/ Limit Generation of Explosive Dust Clouds

Most commonly the dust explosion hazard is combated by

adding preventive and mitigatory measures to an existing process. However, the technical measures adopted are often expensive, and safety procedures may fail.

Inherent safety is an alternative approach. It implies that the process itself be designed in such a way that no explosion hazard exists. Kletz [52], the "father" of the inherently-safer-process design concept, outlined its basic philosophy and recommended the use of it whenever feasible. In the context of preventing and mitigating dust explosions inherently safer process design could include use of production, treatment, transportation, and storage operations where dust cloud generation is kept at a minimum. One example is the use of mass flow silos and hoppers instead of funnel flow types. Eckhoff [53] emphasized the importance of knowing powder science and technology when striving for inherently safer process design in industries having a dust explosion hazard. Amyotte and Khan [54] proposed a framework for directing the concept of inherently safer process design specifically toward reducing the dust explosion hazard in industry. Recently Amyotte et al. [55] described in greater detail how the inherent safety principles of minimization, substitution, moderation, and simplification can be implemented in practice to prevent and mitigate accidental dust explosions in process plant. Hopefully such initiatives will promote further work in this important area.

Inerting by Adding Inert Gas

Explosive dust clouds can be made inert by mixing the air with an inert gas such as nitrogen or carbon dioxide to a level at which the dust cloud can no longer propagate a self-sustained flame. Some further insight has been gained during the last two decades. Wilén et al. [56] found that the limiting oxygen concentration (LOC) for inerting of biomass dust clouds increased with increasing initial pressure of the cloud in the range 5 to 18 bar. This is opposite to the decrease of LOC with increasing initial pressure found earlier for clouds of coal dust. As would be expected, Schwenzfeuer et al. [57] found that LOC for ignition of dust clouds by electrostatic discharges, or metal sparks from mechanical impact, was significantly higher than the conservative limit determined in standard tests, using a very strong pyrotechnical ignition source.

Whilst adding nitrogen to the air can prevent dust explosions, it may introduce a suffocation risk. However, it has been shown that addition of a few vol. % of CO_2 to the nitrogen/air mixture reduces the critical oxygen threshold for suffocation considerably . A gas mixture for inerting utilizing this effect was presented by Dansk Fire Eater A/S [58].

Keeping the Dust Concentration below LEL

In principle, keeping the concentration of dust in the cloud below the lower explosive limit (LEL) is a means of maintaining dust clouds nonexplosive. However, with a few exceptions the method has limited applicability in practice in the process industries. Mittal [59] discussed various mathematical models for calculating minimum explosive concentrations of dust clouds.

PREVENTING IGNITION SOURCES

Smouldering Layers, Deposits, and Nests

Can metal particle sparks from single accidental impacts initiate combustion in dust layers/deposits? Hesby [60] found that the number of sparks from single accidental impacts of steel objects is far too low to be able to cause ignition of the layers of a selection of organic dusts, including tobacco. Gummer and Lunn [61] found that, in general, smouldering nests were poor ignition sources for most dust clouds, whereas flaming nests caused ignition more readily. More work is needed to clarify both the conditions under which smouldering or flaming nests of various materials are generated in industrial plant and the circumstances under which such nests will ignite explosive clouds of various dusts.

Krause and Hensel [62] presented a numerical method by which nonsteady temperature fields in dust deposits can be computed. This enables numerical analysis of a number of practical cases of self heating/self ignition that cannot be analyzed using the classical thermal explosion theory of Frank-Kamenetzki. Krause and Schmidt [63] studied experimentally critical thermal conditions that may lead to initiation of smouldering processes, or to further development of such processes, once initiated.

Hot Surfaces

In the past, the minimum hot-surface temperature for ignition of a dust cloud has often been regarded as if it were a universal constant for a given cloud. Consequently, results from small-scale laboratory tests were often applied directly in design of large-scale industrial plant. However, minimum hot-surface ignition temperatures of dust clouds vary significantly with scale as well as with the geometry of the hot surface in relation to the dust cloud. There is a need for both a more differentiated basic understanding and a more differentiated testing approach. Development of numerical models for dynamic simulation of hot-surface ignition processes encountered in practice is foreseen.

Electric/Electrostatic Discharges between Two Metal Electrodes

Electric and electrostatic discharges between two metal electrodes can be generated in a number of ways, for example, in switches, by failures in electric circuits and by discharge of static electricity. The parameters influencing the minimum energy required for igniting a dust cloud by an electric spark include voltage and current characteristics across the spark gap, spark gap geometry, and electrode material as well as all the dust cloud parameters. The latter include particle material and particle size/shape distributions, dust moisture content, dust concentration, and the dynamic state of the dust cloud with respect to the spark gap. Minimum ignition energies (MIEs) of clouds of a given dust material decrease strongly with the fineness of the dust.

Eckhoff [64] discussed the influence of dust fineness on MIE of ferro-alloys dusts. In the past dust fineness was often specified just as a mass percentage finer than an arbitrary size, for example, 74 µm or 63 µm, without any specification of the distribution of particle sizes below these limits. This complicates the analysis of published experimental data, and more systematic research is needed to clarify the exact influence of particle size. In the case of metal alloys the most hazardous components may sometimes accumulate in the fine tail of the particle size distribution (e.g., Mg in MgFeSi), and special investigations are required.

Lorenz and Schiebler [65] presented the results from a comprehensive, detailed experimental and theoretical investigation of the energy transfer processes taking place during an electrostatic spark discharge. The temperature and pressure development in the spark channel during its formation and subsequent expansion were investigated. This also included cooling of the channel by thermal radiation. The dependence of the ability of a given discharged electrical energy to ignite a dust cloud on these basic physical spark characteristics was emphasized.

Randeberg and Eckhoff [66] investigated an alternative method for measuring MIEs of explosive dust clouds, which may be in better accordance with accidental electrostatic spark ignition in industrial plant. They used the transient dust cloud itself to initiate spark breakdown between a pair of electrodes preset at a high voltage somewhat below the breakdown voltage in dust-free air. The MIEs obtained were of the same order as those obtained using the conventional synchronized-spark methods. The lower spark energy limit for apparatuses commonly used for determining MIEs of dust clouds has been 1–3 mJ, but more recently Randeberg et al. [67] presented a new test method that in principle permits MIE determination for dust clouds, using synchronized sparks, down to the order of 0.1 mJ. However, as pointed out by Eckhoff et al. [68], due to the design of the spark generator used, the spark energies quoted by Randeberg et al. were in fact considerably smaller than the real energies in their experiments. Work is currently being conducted at the University of Bergen to eliminate this problem. Whether new MIE measurements using the improved spark discharge circuit will necessitate adjustment of the very low MIEs for some dust clouds reported by Randeberg and Eckhoff [69] and Eckhoff and Randeberg [70] remains to be seen. Recently Wu et al. [71], using a measurement system with a lower spark energy limit of 1 mJ, reported that clouds in air of a number of very fine titanium and iron powders could be readily ignited by a spark of 1 mJ energy, which means that their MIEs were in fact significantly lower than this value.

Baudry et al. [72], Nifuku et al. [73], and Marmo and Caravello [74] measured MIEs of clouds in air of various types of dust (aluminium with various contents of oxide, aluminium and magnesium dusts from shedding processes, and nylon fibres, resp.).

Electrostatic One-Electrode Discharges

With regard to the even more complex one-electrode types of electrostatic discharges (corona, brush, propagating brush, etc.), valuable experimental insight has been gained during the last years. For example, the issue of whether brush discharges can ignite dust clouds in air was revisited experimentally by Larsen et al. [75], who were able to ignite clouds of fine sulphur dust in oxygen-enriched air by true brush discharges. However, ignition in air only was never observed. Because of the very low MIE of the sulphur dust used, this indicates that ignition of even the most sensitive dust clouds by brush discharges in air is unlikely.

Glowing/Burning Particles

Ignition of dust clouds by small burning metal particles called impact sparks or metal particle sparks is a complex process. Such sparks are generated by single, fast impacts between solid materials, of which one is a metal. So far practically useful theories, describing both impact and ignition, do not seem to be within sight. Such theories must comprise several complex subprocesses. The first is the generation and initial heating of the metal particle by the impact. The second is the ignition of the flying hot particle and the subsequent burning process. The third is the heat transfer to the dust cloud, which ultimately determines whether ignition occurs or not.

Electrical Apparatus

The present situation internationally concerning standards for electrical apparatus for use in areas containing combustible dust is confusing, as discussed by Eckhoff [76, 77]. The International Electrotechnical Commission (IEC) has decided to base its development on the European Union "Atex" philosophy. However, the European Union Atex 94/9/EC Directive does not distinguish adequately between combustible dusts and combustible gases/vapors. This has given rise to undue alignment of a series of new IEC standards for electrical apparatus for combustible dusts with established standards for gases/vapors. The current European Union Atex 1999/92/EC Directive also lacks the required distinction between gases and dusts, which gives rise to problems with area classification.

Other Ignition Sources

Proust [78] determined experimentally the minimum laser beam power required for igniting dust clouds by the heat absorbed by a solid target heated by the laser beam. The variable parameters included the laser beam diameter, the duration of the irradiation, the target material (combustible/noncombustible), and the type of dust. Initiation of dust explosions by shock waves has been studied by several workers, including Wolanski [2] and Klemens et al. [79].

PROTECTIVE/MITIGATORY MEASURES

Full Confinement

The applicability of the concept is limited because of high equipment costs. However, the method is used in some special cases, for example, when the powder/dust is highly toxic, and completely reliable confinement is absolutely necessary. Whereas current experimental methods allow accurate prediction of maximum attainable explosion pressures in simple vessels with point source ignition, design of pressure resistant process equipment may not be straightforward. The use of "finite element" computation methods to achieve improved design seems to be inceasing.

Explosion Isolation

The objective of explosion isolation is to prevent dust explosions from spreading from the primary explosion site to other process units, workrooms, and so forth. Basic understanding of flame propagation and pressure build-up in coupled process equipment ("interconnected vessels") is required for specification of performance criteria of various types of active and passive isolation equipment. van Wingerden et al. [80] reported on dust explosion experiments in a system of two vented vessels connected by a duct. Holbrow et al. [81, 82] summarized the results from extensive similar experiments in the UK and presented coherent quantitative guidance for design of interconnected process equipment based on containment and explosion venting. Vogl and Radandt [83, 84] presented results from a comprehensive experimental program in Germany on propagation of dust explosions in interconnected process systems. Various passive and active techniques for interrupting explosions in pipelines have been developed, but there is room for further improvement.

Partial Inerting

This is a relatively new concept for mitigating dust explosions, which deserves some attention. The idea is that as the oxygen content in the atmosphere is reduced, there is a systematic decrease of both ignition sensitivity and combustion rate of the dust cloud. In many cases the explosion hazard may be reduced markedly by only a moderate reduction of the oxygen content. Glor and Schwenzfeuer [85] confirmed experimentally that even modest reductions of the oxygen content can increase the minimum ignition energies of dust clouds substantially. Devlikanov et al. [86] found that K_{st} was a linear function of the percentage of oxygen in the gas phase (mixture of nitrogen and oxygen). Conde Lázaro and García Torrent [87] carried out a series of partial inerting experiments at 12 bar initial pressure, in a demonstration pulverized coal power plant. Eckhoff [88] called for more extensive use of partial inerting in industrial dust explosion protection at large.

Explosion Venting

This is probably the most widely used method for mitigating dust explosions. In spite of extensive research and development, dust explosion venting remains a complex and in part controversial subject. The key issue is vent area sizing.

Tamanini and Valiulis [89] presented an improved version of the VDI (Germany) and NFPA (USA) guidelines for sizing of dust explosion vents. The improvement was achieved by systematizing the data in the context of a simplified physical model of the vented explosion. A similar contribution was made by Ural [90]. The new CEN [91] standard for design of dust explosion venting systems in principle opens up for a differentiated approach to vent sizing, which accounts for the variations in dust cloud structures encountered in practice in industry. In most practical cases this will result in more liberal vent area requirements than those of some previous rigorous standards.

Other aspects of explosion venting studied more recently include the influence of the inertia and specific design of the vent cover on the gas dynamics of the venting process. A further dimension of complexity is added to the venting problem if the initial pressure (and/or temperature) deviates from atmospheric.

Results from venting of dust explosions in air at elevated initial pressure were reported by Siwek et al. [92].

In dust explosion venting, maintaining the integrity of the enclosure is not the only concern. Venting implies that both blast waves and flames are emitted into the surroundings, and this may present a hazard, depending on the size of the emitted flame and the magnitude of the blast wave. Several workers, including Forcier and Zalosh [93], Holbrow et al. [94] and Harmanny [95], investigated various aspects of this problem. Various methods have been developed for eliminating hazardous effects of flames from vent openings. Li et al. [96] and Emde and Penno [97] discussed further aspects of the Q-pipe for dust and flame-free venting. The influence of vent ducts on the maximum explosion pressure in the vented vessel has been studied experimentally by several workers including Ural [98] and Lunn [99]. Tamanini and Valiulis [100] presented a new theoretical approach for predicting the resultant reaction impulse acting on a process structure during a vented explosion.

Venting of industrial buildings requires special considerations. An overview was given by Crowhurst [101]. Höppner [102] discussed the design of dust explosion venting arrangements for rooms/buildings of volumes >5000m^3, with walls that can only withstand overpressures less than 0.2 bar. In case of a dust explosion, only part of such large volumes will be filled with explosive dust cloud.

Tamanini [103] summarized his valuable effort of correlating existing experimental dust explosion venting data by applying the classical method of dimensional analysis. It is regrettable that this important work was not included in the recent European Union guideline for design dust explosion venting arrangements, CEN [91], but it has been included in recent US standards (NFPA). Silvestrini et al. [104] developed correlations for flame speed and explosion overpressure for dust explosions inside industrial enclosures with the aim to provide a simple tool for sizing dust explosion vents.

However, in view of the different turbulence levels, degrees of dust dispersion, and distributions of dust concentrations encountered in industry, the need for a more complete differentiated approach to assessment of vent area requirements has gained general acceptance. As indicated by Skjold [34], the numerical code DESC may become a useful tool to meet this need.

Automatic Explosion Suppression

This active method for dust explosion mitigation is comparatively complex and expensive. It is therefore used when simpler and less expensive methods are inadequate. The method has been in use for many decades, and significant progress has been made during the last decade. For example, Moore [105] and Chatrathi and Going [106] evaluated the suitability of various suppressants, and Tyldesley [107] reported that superheated water can in some circumstances be an effective candidate. Moore and Siwek [108] summarized their extensive multiyear experimental work on suppression of dust explosions, whereas Chatrathi and Going [109] gave an overview of current technology and philosophy for implementing automatic explosion suppression systems in practice. The influence of elevated initial temperature of the explosive dust cloud on the efficacy of an automatic explosion suppression system was studied by Brehm [110].

The European standardization organization CEN [111] has produced a draft standard for design of explosion suppression systems, which seems to open up for greater flexibility than the traditional, mostly very conservative approach. Hence, if the turbulence level and/or degree of homogeneity of the cloud of a given dust in the actual process situation are lower than those produced by the rather conservative traditional standard VDI-method of dust cloud generation, this can be accounted for in the design of the suppression system.

Comprehensive numerical models of the complex explosion suppression process, based on computational fluid dynamics (CFD), are likely to be the future tool for design of optimal explosion suppression systems. Again the code DESC should be mentioned as a promising candidate. Morgan [112] assessed the suitability of commercially available CFD software for modeling the types of flows encountered in explosion suppression processes. Using results from his model simulations, he was able to design a novel suppressant injection nozzle, which was shown to be more effective than standard nozzles currently used.

Design of Process Equipment for Specific Internal Explosion Loads

This problem is also addressed in 6.1 above and is a central concern also when designing explosion venting systems (6.4) and systems for automatic explosion suppression (6.5). Harmanny [113] presented a new equation for predicting the duration of vented dust explosions in enclosures of volumes from 10 to 60m^3. This is a useful tool for evaluating whether static pressure considerations or impulse considerations apply when predicting the response of the enclosure structure to the explosion load. Harmanny [114, 115] revisited the problem of assessing the structural response of a given process equipment and buildings to explosion loads. With regard to dust explosions in the process industries, he concluded that most often they are sufficiently slow for the load to be regarded as quasistatic. However, there are certain cases where dynamic effects play a significant role. Comprehensive finite-element-based computer codes for determining detailed stress/strain analyses of complex structures exposed to defined static and dynamic loads have been available for some time. It is foreseen that the use of such tools in assessing the explosion strength of complex process equipment will increase in the years to come. The concept of pressure-shock-resistant design should be developed further to facilitate cost effective equipment design. Li et al. [116] compared elastic and plastic structural responses of a simple mechanical structure determined experimentally with predictions from using a computational finite-element approach.

Preventing Secondary Explosions outside

Process Equipment

This remains a most important issue in all efforts to fight the dust explosion hazard. Adequate housekeeping is an essential means of achieving this aim. However, there are still questions to be answered concerning the level of cleanliness required. More research is needed for assessment of the maximum acceptable mass of deposited dust per unit area of surface for preventing secondary dust flame propagation under various conditions. Cybulski et al. [117] showed that comparatively weak secondary dust explosions in short narrow tunnels in grain elevators can be extinguished by properly designed, actively triggered water barriers. They also showed that, under the conditions prevailing, the possibility of flame penetration into blind gallery branches was small. This kind of work may also be of relevance to the analysis of flame propagation in large industrial systems, for example, in grain storage and handling plants.

OTHER FACTORS INFLUENCING THE DUST EXPLOSION RISK

Explosion Risk Management

Barth [118] emphasized the importance of companies establishing systems for explosion risk management control to ensure effective, long-lasting explosion protection of process plant. Hesener et al. [119], with reference to the pharmaceutical industry, underlined the need for having adequate systems for explosion risk management and control even in small and medium size plants. van der Voort et al. [120] presented a quantitative risk assessment tool for the external safety of industrial plants with a dust explosion hazard.

Cost/Benefit in Dust Explosion Prevention and Mitigation

Alfert [121] addressed the bottom-line costs of various dust explosion protection systems on the market. Janssens [122] pointed out that the investments required to achieve proper prevention and control of the explosion hazard in a given plant are not necessarily excessive. By combining thorough knowledge of the processes to be protected, with knowledge of relevant ignition and flame propagation phenomena, and principles and technologies available for explosion control, good solutions can be obtained at an acceptable cost.

Education and Training

High safety levels in the process industries cannot be established once and for all by a single all-out effort. Deterioration results if the high level once attained is not actively secured by continuous maintenance and renewal. This applies both to technology and human factors. Education and training, from short practical training courses to in-depth long-term education, play a key role in the continuous maintenance and renewal process. Universities and colleges have responded to this challenge by establishing study courses on a wide range of process safety aspects. Relevant topics include reliability and risk analysis, the physics, chemistry, and technology of processes and hazards, and means of accident prevention and mitigation. Much emphasis has been put on methods of reliability and risk analysis, which are indeed very important. However, it is sometimes felt by the process industry itself that education in the "hard" aspects, that is, the physics, chemistry, and technology of processes and process hazards, has been somewhat left behind. This situation presents a special challenge to universities and colleges.

PERSPECTIVES FOR THE FUTURE

- The approaches taken in dust explosion prevention and mitigation in the process industries will become steadily less dogmatic and more tailored and differentiated in the years to come. Industry will strive for steadily more cost effective safety measures.
- Therefore, substantial progress is foreseen in mathematical CFD-based modeling of dust cloud generation and flame propagation processes in dust clouds. It is anticipated that such models will gradually replace conventional empirical equations and graphs as design tools for tailored systems for explosion isolation, explosion venting, and automatic explosion suppression and for evaluating consequences of secondary dust explosions. However, extensive experimental validation of the numerical models is absolutely necessary.
- Evaluation of potential ignition sources will also become more detailed and differentiated, in accordance with reality. Further development of mathematical models capable of predicting ignition of dust clouds and layers by self-heating/smouldering, hot surfaces, various electrical and electrostatic sparks/discharges, metal sparks, and so forth, is foreseen.
- Combined protective solutions, for example, partial inerting in combination with venting, or venting combined with automatic suppression, are likely to be used to an increasing extent.
- The need for inherently, safer design of processes for production, treatment and handling of combustible powders/dusts is expected to be become more and more accepted. To achieve this, knowledge and systematic use of powder/particle science and technology is a basic requirement.
- High-quality training/education, ranging from short courses of a few days to extensive university studies, will continue to be essential for minimizing the hazards in the process industries, including minimizing the risk of dust explosions.

REFERENCES

1. J. H. S. Lee, F. Zhang, and R. Knystautas, "Propagation mechanisms of combustion waves in dust-air mixtures," Powder Technology, vol. 71, no. 2, pp. 153–162, 1992.
2. P. Wolanski, Deflagration, Detonation and Combustion of Dust Mixtures, American Institute of Aeronautics and Astronautics, New York, NY, USA, 1990.
3. R. K. Eckhoff, "Generation, ignition, combustion and explosion of sprays and mists of flammable liquids in air. A Literature Survey," Tech. Rep. CMI–91–A25014, Christian Michelsen Institute, Fantoft, Norway, 1991.
4. R. K. Eckhoff, Dust Explosions in the Process Industries, Gulf Professional Publishing/Elsevier, Boston, Mass, USA, 3rd edition, 2003.
5. S. M. Frolov, K. A. Avdeev, and F. S. Frolov, "Effect of transient heat transfer on ignition of solid particles," Journal of Loss Prevention in the Process Industries, vol. 20, no. 4–6, pp. 310–316, 2007.
6. A. V. Fedorov and A. V. Shulgin, "About stability of the ignition process of small solid particle," Journal of Loss Prevention in the Process Industries, vol. 20, no. 4–6, pp. 317–321, 2007.
7. A. E. Dahoe, K. Hanjalic, and B. Scarlett, "Determination of the laminar burning velocity and the Markstein length of powder-air flames," Powder Technology, vol. 122, no. 2-3, pp. 222–238, 2002.
8. I. O. Moen, J. H. S. Lee, and B. H. Hjertager, "Pressure development due to turbulent flame propagation in large-scale methane—air explosions," Combustion and Flame, vol. 47, pp. 31–52, 1982.
9. B. H. Hjertager, K. Fuhre, and M. Bjoerkhaug, "Gas explosion experiments in 1:33 and 1:5 scale offshore separator and compressor modules using stoichiometric homogeneous fuel/air mixtures," Journal of Loss Prevention in the Process Industries, vol. 1, no. 4, pp. 197–220, 1988.

10. J. R. Bakke and K. van Wingerden, Guidance for Designing Offshore Modules Evolving from Gas Explosion Research, Society of Petroleum Engineers, Richardson, Tex, USA, 1992.
11. F. Rzal-Rebière and B. Veyssière, "Propagation mechanisms of starch particles-air flames," in Proceedings of the 6th International Colloquium on Dust Explosions, D. Xufan and P. Wolanski, Eds., pp. 186–200, Shenyang, China, August-September 1994.
12. P. Mitgau, Einfluss der Turbulenzlänge und der Schwankungsgeschwindichkeit auf die Verbrennungsgeschwindigkeit von aerosolen, vol. 14, Max-Planck-Institut Für Strömungsforschung, Göttingen, Germany, 1996.
13. P. Mitgau, H. Gg. Wagner, and R. Klemens, "Einfluss der Turbulenzlänge und der Schwankungsgeschwindichkeit auf die Flammengeschwindigkeit von Stäuben," in Feuerungstechnik, Kaleidoskop aus aktueller Forschung und Entwicklung. Geburtstag, Festschrift an Prof. Wolfgang Leuckel zu seinem 65, pp. 17–45, Engler-Bunte-Institut, Bereich Feuerungstechnik, Universität Karlsruhe (TH), Geburtstag, Germany, 1997.
14. K. L. Cashdollar and I. A. Zlochower, "Explosion temperatures and pressures of metals and other elemental dust clouds," Journal of Loss Prevention in the Process Industries, vol. 20, no. 4–6, pp. 337–348, 2007.
15. L. Kjäldman, "Numerical flow simulation of dust deflagrations," Powder Technology, vol. 73, no. 1, p. 100, 1992.
16. M. Rose, P. Roth, S. M. Frolov, and M. G. Neuhaus, "Modelling of turbulent gas/particle combustion by a Lagrangian PDF method," Combustion Science and Technology, vol. 149, no. 1, pp. 95–113, 1999.
17. N. N. Smirnov, V. F. Nikitin, and J. C. Legros, "Ignition and combustion of turbulized dust-air mixtures," Combustion and Flame, vol. 123, no. 1-2, pp. 46–67, 2000.
18. U. Bielert and M. Sichel, "Numerische simulation von staubexplosionen in pneumatischen saug-flug-förderanlagen,"

VDI-Berichte, no. 1601, pp. 449–472, 2001.
19. K. Wörsdörfer, M. Sippel, J. Fuisting, and A. Kneer, "Möglichkeiten des Einsatzes numerischer Methoden im Explosionsschutz," VDI-Berichte, no. 1601, pp. 437–447, 2001.
20. V. P. Korobeinikov, I. V. Semenov, I. S. Menshov, R. Klemens, P. Wolanski, and P. Kosinski, "Modelling of flow and combustion behind shock waves propagating along dust layers in long ducts," Journal de Physique IV, vol. 12, no. 7, pp. Pr7/113–Pr7/119, 2002.
21. S. Zhong, A. Teodorczyk, X. Deng, and J. Dang, "Modeling and simulation of coal dust explosions," Journal de Physique IV, vol. 12, no. 7, pp. Pr7/141–Pr7/147, 2002.
22. P. Kosinski, R. Klemens, and P. Wolanski, "Potential of mathematical modelling in large-scale dust explosions," Journal de Physique IV, vol. 12, no. 7, pp. Pr7/125–Pr7/132, 2002.
23. A. Di Benedetto and P. Russo, "Thermo-kinetic modelling of dust explosions," Journal of Loss Prevention in the Process Industries, vol. 20, no. 4–6, pp. 303–309, 2007.
24. K. van Wingerden, B. J. Arntzen, and P. Kosi ski, "Modelling of dust explosions," VDI-Berichte, no. 1601, pp. 411–421, 2001.
25. B. J. Arntzen, H. C. Salvesen, H. F. Nordhaug, I. E. Storvik, and O. R. Hansen, "CFD-modelling of oil mist and dust explosion experiments," in Proceedings of the 4th International Seminar on Fire and Explosion Hazards, pp. 601–608, Londonderry, UK, September 2003.
26. O. R. Hansen, T. Skjold, and B. J. Arntzen, "DESC—a CFD tool for dust explosions," in Proceedings of the 3rd International ESMG Symposium on Process Safety and Industrial Explosion Protection, Nürnberg, Germany, March 2004.
27. R. Siwek, K. Wingerden, O. R. van Hansen, et al., "Dust explosion venting and suppression of conventional spray dryers," in Proceedings of the 11th International Symposium

Loss Prevention and Safety Promotion in the Process Industries, Praha, Czech Republic, May-June 2004.

28. T. Skjold, B. J. Arntzen, O. R. van Hansen, I. Storvik, and R. K. Eckhoff, "Simulation of dust explosions in complex geometries with experimental input from standardized tests," in Proceedings of the 5th International Symposium on Hazards, Prevention and Mitigation of Industrial Explosions (ISHPMIE ‹04), Krakow, Poland, October 2004.

29. T. Skjold, B. J. Arntzen, O. R. van Hansen, O. J. Taraldset, I. Storvik, and R. K. Eckhoff, "Simulating dust explosions with the first version of DESC," in Proceedings of the Symposium on Hazards XVIII: Process Safety—Shearing Best Practice, IChemE NW Branch Symposium, UMIST, Manchester, UK, November 2004.

30. T. Skjold, B. J. Arntzen, O. R. van Hansen, O. J. Taraldset, I. E. Storvik, and R. K. Eckhoff, "Simulating dust explosions with the first version of DESC," Process Safety and Environmental Protection, vol. 83, no. 2, pp. 151–160, 2005

31. T. Skjold, B. J. Arntzen, O. R. van Hansen, I. E. Storvik, and R. K. Eckhoff, "Simulation of dust explosions in complex geometries with experimental input from standardized tests," Journal of Loss Prevention in the Process Industries, vol. 19, no. 2-3, pp. 210–217, 2006.

32. T. Skjold, "Review of the DESC project," in Proceedings of the 6th International Symposium on Hazards, Prevention, and Mitigation of Industrial Explosions (ISHPMIE ‹06), pp. 1–21, Halifax, Canada, August-September 2006, (Key-note paper).

33. T. Skjold, R. K. Eckhoff, B. J. Arntzen, et al., "Simplified modelling of explosion propagation by dust lifting in coal mines," in Proceedings of the 5th Intenational Seminar on Fire and Explosion Hazards, The University of Edinburgh, Scotland, UK, April, 2007.

34. T. Skjold, "Review of the DESC project," Journal of Loss Prevention in the Process Industries, vol. 20, no. 4–6, pp. 291–302, 2007.

35. R. Weber, "Preisgekrönte Abhandlung über die Ursachen von Explosionen und Bränden in Mühlen, sowie über die Sicherheitsmassregein zur Verhütung derselben,"Verhandlungen des Vereins zur Beförderung des Gewerbe- fleißes, pp. 83–103, 1878
36. F. Hauert, A. Vogl, and S. Radandt, "Measurement of turbulence and dust concentration in silos and vessels," in Proceedings of the 6th International Colloquium on Dust Explosions, D. Xufan and P. Wolanski, Eds., pp. 71–80, Shenyang, China, August-September 1994.
37. P. Kosinski, R. Klemens, P. Wolanski, V. P. Korobeinikov, V. V. Markov, and I. S. Men›shov, "Dust-air mixtures spreading in branched ducts," in Proceedings of the 18th International Colloquium Dynamics Exploration & Reaction System, Seattle, Wash, USA, 2001.
38. P. Kosinski and A. C. Hoffmann, "Modelling of dust lifting using the Lagrangian approach," International Journal of Multiphase Flow, vol. 31, no. 10-11, pp. 1097–1115, 2005.
39. P. Kosinski, A. C. Hoffmann, and R. Klemens, "Dust lifting behind shock waves: comparison of two modelling techniques," Chemical Engineering Science, vol. 60, no. 19, pp. 5219–5230, 2005.
40. C. G. Ilea, P. Kosinski, and A. C. Hoffmann, "Three-dimensional simulation of a dust lifting process with varying parameters," International Journal of Multiphase Flow, vol. 34, no. 9, pp. 869–878, 2008.
41. K. Lebecki, J. Sliz, Z. Dyduch, and P. Wolanski, Critical Dust Layer Thickness for Combustion of Grain Dust, American Institute of Aeronautics and Astronautics, New York, NY, USA, 1990.
42. C. W. Kauffman, M. Sichel, and P. Wolanski, "Research on dust explosions at the University of Michigan," Powder Technology, vol. 71, no. 2, pp. 119–134, 1992.
43. P. J. Austin, F. Girodroux, Y. C. Li, C. G. Alexander, C. W.

Kauffman, and M. Sichel, "Recent progress in the study of dust combustion phenomena at the University of Michigan," in Proceedings of the 5th International Colloquium on Dust Explosions, pp. 211–214, Pultusk, Poland, April 1993.

44. V. M. Boiko and S. V. Poplavski, "On the effect of particle concentration on acceleration of a dusty cloud behind a shock wave," in Proceedings of the 7th International Colloquium on Dust Explosions, GexCon AS, Bergen, Norway, June 1996.

45. R. Klemens, P. Kosinski, and P. Oleszczak, "Mathematical modelling of dust layer dispersion by rarefaction waves," Archivum Combustionis, vol. 22, no. 1-2, pp. 3–12, 2002.

46. A. V. Fedorov and Yu. A. Gosteev, "Quantitative description of lifting and ignition of organic fuel dusts in shock waves," Journal de Physique IV, vol. 12, no. 7, pp. Pr7/89–Pr7/95, 2002.

47. A. V. Fedorov and N. N. Fedorova, "Numerical simulations of dust lifting under the action of shock wave propagating along the near-wall layer," Journal de Physique IV, vol. 12, no. 7, pp. Pr7/97–Pr7/104, 2002.

48. F. Tamanini and E. A. Ural, "FMRC studies of parameters affecting the propagation of dust explosions," Powder Technology, vol. 71, no. 2, pp. 135–151, 1992.

49. J. A. H. de Jong, A. C. Hoffmann, and H. J. Finkers, "Properly determine powder flowability to maximize plant output," Chemical Engineering Progress, vol. 95, no. 4, pp. 25–34, 1999.

50. N.O. Breum, "The rotating drum dustiness tester: variability in dustiness in relation to sample mass, testing time, and surface adhesion," Annals of Occupational Hygiene, vol. 43, no. 8, pp. 557–566, 1999.

51. D. Dahmann and K. Möcklinghoff, "Das Staubungsverhalten quarzfeinstaubhaltige Produkte," Gefahrstoffe- Reinhaltung der Luft, vol. 60, pp. 213–215, 2000.

52. T. Kletz, "Inherently safer design: avoidance better than control," in Proceedings of the 3rd World Seminar on the Explosion Phenomenon and on the Application of Explosion Protection Techniques in Practice, Flanders Expo, Gent, Belgium, February 1999.
53. R. K. Eckhoff, "Understanding dust explosions. The role of powder science and technology," Journal of Loss Prevention in the Process Industries, vol. 22, no. 1, pp. 105–116, 2009.
54. P. R. Amyotte and F. I. Khan, "An inherent safety framework for dust explosion prevention and mitigation," Journal de Physique IV, vol. 12, no. 7, pp. Pr7/189–Pr7/196, 2002.
55. P. R. Amyotte, M. J. Pegg, and F. I. Khan, "Application of inherent safety principles to dust explosion prevention and mitigation," Process Safety and Environmental Protection, vol. 87, no. 1, pp. 35–39, 2009.
56. C. Wilén, A. Rautalin, J. García-Torrent, and E. Conde-Lázaro, "Inerting biomass dust explosions under hyperbaric working conditions," Fuel, vol. 77, no. 9-10, pp. 1089–1092, 1998.
57. K. Schwenzfeuer, M. Glor, and A. Gitzi, "Relation between ignition energy and limiting oxygen concentrations for powders," in Proceedings of the 10th International Symposium Loss Prevention and Safety Promotion in the Process Industries, H. J. Pasman, O. Fredholm, and A. Jacobsson, Eds., pp. 909–916, Elsevier, Stocholm, Sweden, June 2001.
58. Dansk Fire Eater A/S, "INERGEN. Anlœgsbeskrivelse & Design," Report, Dansk Fire Eater A/S, Holte, Denmark, 1992.
59. M. Mittal, "Mathematical models for minimum explosible concentration of dusts," in Proceedings of the 5th International Colloquium on Dust Explosions, pp. 247–256, Pultusk, Poland, April 1993.
60. I. Hesby, Ignition of dust layers by metal particle sparks, M.Sc. thesis, Department of Physics, University of Bergen, Bergen, Norway, 2000.
61. J. Gummer and G. Lunn, "Ignitions of explosive dust clouds by smouldering and flaming agglomerates," Journal of Loss

Prevention in the Process Industries, vol. 16, no. 1, pp. 27–32, 2003.

62. U. Krause and W. Hensel, "Zündgefahren lagernder Staubschüttungen—Neue Hilfsmittel für ihre Bewertung," VDI-Berichte, no. 1272, pp. 183–201, 1996.

63. U. Krause and M. Schmidt, "Untersuchungen zur Zündung und Ausbreitung von Schwelbränden in Stäuben und Schüttgütern," VDI-Berichte, no. 1601, pp. 397–410, 2001.

64. R. K. Eckhoff, "Dust explosion hazards in the ferro-alloys industry," in Proceedings of the 52nd Electric Furnace Conference, pp. 283–302, Iron and Steel Society, Nashville, Tenn, USA, November 1995.

65. D. Lorenz and H. Schiebler, "Optische Temperaturmessung an Entladungsfunken im Hinblick auf deren Zündwirksamkeit bei Staubexplosionen," VDI-Berichte, no. 1601, pp. 653–667, 2001.

66. E. Randeberg and R. K. Eckhoff, "Initiation of dust explosions by electric spark discharges triggered by the explosive dust cloud itself," in Proceedings of the 5th International Symposium on Hazards, Prevention and Mitigation of Industrial Explosions (ISHPMIE ‹04), Krakow, Poland, October 2004.

67. E. Randeberg, W. Olsen, and R. K. Eckhoff, "A new method for generation of synchronised capacitive sparks of low energy," Journal of Electrostatics, vol. 64, no. 3-4, pp. 263–272, 2006.

68. R. K. Eckhoff, W. Olsen, and O. Kleppa, "Influence of spark discharge duration on the minimum ignition energy of premixed propane/air," in Proceedings of the 7th International Symposium on Hazards, Prevention, and Mitigation of Industrial Explosions (ISHPMIE ‹08), vol. 1, pp. 44–53, St. Petersburg, Russia, July 2008.

69. E. Randeberg and R. K. Eckhoff, "Measurement of minimum ignition energies of dust clouds in the <1 mJ region," Journal of Hazardous Materials, vol. 140, no. 1-2, pp. 237–244, 2007.

70. R. K. Eckhoff and E. Randeberg, "Electrostatic spark ignition of

sensitive dust clouds of MIE<1 mJ," Journal of Loss Prevention in the Process Industries, vol. 20, no. 4–6, pp. 396–401, 2007.

71. H.-C. Wu, R.-C. Chang, and H.-C. Hsiao, "Research of minimum ignition energy for nano titanium powder and nano Iron powder," Journal of Loss Prevention in the Process Industries, vol. 22, no. 1, pp. 21–24, 2009.

72. G. Baudry, S. Bernard, and P. Gillard, "Influence of the oxide content on the ignition energies of aluminium powders," Journal of Loss Prevention in the Process Industries, vol. 20, no. 4–6, pp. 330–336, 2007.

73. M. Nifuku, S. Koyanaka, H. Ohya, et al., "Ignitability characteristics of aluminium and magnesium dusts that are generated during the shredding of post-consumer wastes,"Journal of Loss Prevention in the Process Industries, vol. 20, no. 4–6, pp. 322–329, 2007.

74. L. Marmo and D. Cavallero, "Minimum ignition energy of nylon fibres," Journal of Loss Prevention in the Process Industries, vol. 21, no. 5, pp. 512–517, 2008.

75. Ø. Larsen, J. H. Hagen, K. van Wingerden, and R. K. Eckhoff, "Ignition of dust clouds by brush discharges in oxygen enriched atmospheres," Gefahrstoffe- Reinhaltung der Luft, vol. 61, no. 3, pp. 85–90, 2001.

76. R. K. Eckhoff, "A critical view on the treatment of combustible powders/dusts in the European 'Atex 100a' and 'Atex 118a' Directives," in Proceedings of the 3rd International ESMG Symposium on Process Safety and Industrial Explosion Protection, Nürnberg, Germany, March 2004.

77. R. K. Eckhoff, "Inadequate treatment of dust explosions and fires in 'ATEX'. A critical view on resulting standards for electrical apparatus," Bulk Solids & Powder Science & Technology. In press.

78. Ch. Proust, "Laser ignition of dust clouds," Journal de Physique IV, vol. 12, no. 7, pp. Pr7/79–Pr7/88, 2002.

79. R. Klemens, P. Wolanski, and J. Klammer, "On unsteady flows of combustible dusty gases caused by a shock wave

propagation," in Proceedings of the 8th International Colloquium on Dust Explosions, pp. 355–363, Safety Consulting Engineers, Schaumburg, Ill, USA, September 1998.

80. K. van Wingerden, G. H. Pedersen, G. H. Teigland, and R. K. Eckhoff, "Violence of dust explosions in integrated systems," in Proceedings of the 28th AIChE Annual Loss Prevention Symposium, American Institute of Chemical Engineers, Atlanta, Ga, USA, April 1995, Session no. 13 on Dust Explosions.

81. P. Holbrow, S. Andrews, and G. A. Lunn, "Dust explosions in interconnected vented vessels," Journal of Loss Prevention in the Process Industries, vol. 9, no. 1, pp. 91–103, 1996.

82. P. Holbrow, G. A. Lunn, and A. Tyldesley, "Dust explosion protection in linked vessels: guidance for containment and venting," Journal of Loss Prevention in the Process Industries, vol. 12, no. 3, pp. 227–234, 1999.

83. A. Vogl and S. Radandt, "Explosionsübertragung durch dünne Rohrleitungen," VDI-Berichte, no. 1601, pp. 575–594, 2001.

84. A. Vogl and S. Radandt, "Explosionsübertragung durch dünne Rohrleitungen," Tech. Rep. 05-9903, Forsch.gesellsch. angew. Systemsicherheit und Arbeitsmedizin, Mannheim, Germany, 2002.

85. M. Glor and K. Schwenzfeuer, "Einfluss der Sauerstoffkonzentration auf die Mindestzündenergie von Staüben," in Dechema Jahrestagung, Wiesbaden, Germany, April 1999.

86. O. Devlikanov, D. K. Kuzmenko, and N. L. Poletaev, "Nitrogen dilution for explosion of nutrient yeast dust/air mixture," Fire Safety Journal, vol. 25, no. 4, p. 373, 1995.

87. E. Conde Lázaro and J. García Torrent, "Experimental research on explosibility at high initial pressures of combustible dusts," Journal of Loss Prevention in the Process Industries, vol. 13, no. 3–5, pp. 221–228, 2000.

88. R. K. Eckhoff, "Partial inerting-an additional degree of freedom in dust explosion protection," Journal of Loss Prevention in the Process Industries, vol. 17, no. 3, pp. 187–193, 2004.
89. F. Tamanini and J. V. Valiulis, "Improved guidelines for the sizing of vents in dust explosions," Journal of Loss Prevention in the Process Industries, vol. 9, no. 1, pp. 105–118, 1996.
90. E. A. Ural, "A simplified development of a unified dust explosion vent sizing formula," in Proceedings of the 35th Annual Loss Prevention Symposium, American Institute of Chemical Engineers, Houston, Tex, USA, April 2001.
91. CEN, "Dust explosion venting protective systems," European Union draft standard prEN 14491 (CEN/TC 305/WG 3/SG 5N, 27 February 2002) prepared by CEN/TC 305 'Potentially explosive atmospheres. Explosion prevention and protection', 2002.
92. R. Siwek, M. Glor, and T. Torreggiani, "Dust explosion venting at elevated initial pressure," in Proceedings of the 7th International Symposium Loss Prevention and Safety Promotion in the Process Industries, pp. 57-1–57-15, SRP-Partners, Roma, Italy, May 1992.
93. T. Forcier and R. Zalosh, "External pressures generated by vented gas and dust explosions," Journal of Loss Prevention in the Process Industries, vol. 13, no. 3–5, pp. 411–417, 2000.
94. P. Holbrow, S. J. Hawksworth, and A. Tyldesley, "Thermal radiation from vented dust explosions," Journal of Loss Prevention in the Process Industries, vol. 13, no. 6, pp. 467–476, 2000.
95. A. Harmanny, "Pressure effects from vented dust explosions," VDI-Berichte, no. 1601, pp. 539–550, 2001.
96. G. Li, X. Deng, W. Liu, et al., "Development of a quenching venting door (QVD)," inProceedings of the 6th International Colloquium on Dust Explosions, D. Xufan and P. Wolanski, Eds., pp. 530–534, Shenyang, China, August-September 1994.

97. A. Emde and B. Penno, "Einbindung der Sauerstoffverdrängung und des Kontraktionseffektes mit angepasstem Wiederstandsbeiwert Zeta bei der Entwicklung neuartiger Quenchvorrichtungen zur Explosionsdruckentlastung innerhalb von Räumen," VDI-Berichte, no. 1272, pp. 645–651, 1996.
98. E. A. Ural, "A simplified method for predicting the effect of ducts connected to explosion vents," Journal of Loss Prevention in the Process Industries, vol. 6, no. 1, pp. 3–10, 1993.
99. G. A. Lunn, "Institution of chemical engineers vent duct method applied to the VDI vent sizing technique," VDI-Berichte, no. 1601, pp. 513–526, 2001.
100. F. Tamanini and J. V. Valiulis, "A correlation for the impulse produced by vented explosions," Journal of Loss Prevention in the Process Industries, vol. 13, no. 3–5, pp. 277–289, 2000.
101. D. Crowhurst, "Explosion protection of industrial buildings," The European Summer School on Dust Explosion Hazards: Their Assessment and Control, organized by IBC Technical Services, in association with BMHB and IELG, Cambridge, UK, 1993.
102. K. Höppner, "Explosionsdruckentlastung von Gebäuden," VDI-Berichte, no. 1272, pp. 327–346, 1996.
103. F. Tamanini, "Dust explosion vent sizing. Current methods and future developments,"Journal de Physique IV, vol. 12, no. 7, pp. Pr7/31–Pr7/44, 2002.
104. M. Silvestrini, B. Genova, and F. J. Leon Trujillo, "Correlations for flame speed and explosion overpressure of dust clouds inside industrial enclosures," Journal of Loss Prevention in the Process Industries, vol. 21, no. 4, pp. 374–392, 2008
105. P. E. Moore, "Suppressants for the control of industrial explosions," Journal of Loss Prevention in the Process Industries, vol. 9, no. 1, pp. 119–123, 1996.
106. K. Chatrathi and J. Going, "Effectiveness of dust explosion suppressants," inProceedings of the 9th International

Symposium on Loss Prevention and Safety Promotion Process Industry, pp. 1008–1017, Barcelona, Spain, May 1998.
107. A. Tyldesley, "Private letter to R. K. Eckhoff," November 1993.
108. P. E. Moore and R. Siwek, "Explosion suppression overview," in Proceedings of the 9th International Symposium Loss Prevention and Safety Promotion in the Process Industries, pp. 745–758, Barcelona, Spain, May 1998.
109. K. Chatrathi and J. Going, "Dust deflagration extinction," Process Safety Progress, vol. 19, no. 3, pp. 146–153, 2000.
110. K. Brehm, "Explosionsunterdrückung bei erhöhter temperatur," VDI-Berichte, no. 1272, pp. 261–272, 1996.
111. CEN, "Explosion suppression systems," European Union draft standard prEN 14373 (CEN/TC 305 WI 00305032, August 2001) prepared by CEN/TC 305 'Potentially explosive atmospheres. Explosion prevention and protection', 2001.
112. A. J. Morgan, The arresting of explosions to minimize environmental damage, Ph.D. thesis, Department of Mechanical Engineering, Brunel University, Uxbridge, UK, 2000.
113. A. Harmanny, "Duration of vented dust explosions," EuropEx Newsletter, vol. 23, pp. 5–9, 1993.
114. A. Harmanny, "Structural aspects related to explosion protection techniques," inProceedings of the 2nd World Seminar on the Explosion Phenomenon and on the Application of Explosion Protection Techniques in Practice, EuropEx, Gent, Belgium, March 1996.
115. A. Harmanny, "Structural aspects related to explosion resistance of process buildings, structures and silos," in Proceedings of the 3rd World Seminar on the Explosion Phenomenon and on the Application of Explosion Protection Techniques in Practice, Flanders Expo, Gent, Belgium, February 1999.
116. G. Li, B.-Z. Chen, X.-F. Deng, and R. K. Eckhoff, "Explosion resistance of a square plate with a square hole," Journal de

Physique IV, vol. 12, no. 7, pp. Pr7/121–Pr7/124, 2002.

117. K. Cybulski, Z. Dyduch, K. Lebecki, and J. Sliz, "Suppression of grain dust explosions with triggered barriers," in Proceedings of the 5th International Colloquium on Dust Explosions, pp. 437–447, Pultusk, Poland, April 1993.

118. U. Barth, "Explosionsgefahren managen—systematisch oder mit system?" VDI-Berichte, no. 1601, pp. 207–223, 2001.

119. U. Hesener, U. Barth, and B. Dyrba, "Erstellung von Explosionsschutzdokumenten anhand von Anlagenbeispielen der pharmazeutischen Industrie," VDI-Berichte, no. 1601, pp. 225–237, 2001.

120. M. M. van der Voort, A. J. J. Klein, M. de Maaijer, A. C. van den Berg, J. R. van Deursen, and N. H. A. Versloot, "A quantitative risk assessment tool for the external safety of industrial plants with a dust explosion hazard," Journal of Loss Prevention in the Process Industries, vol. 20, no. 4–6, pp. 375–386, 2007.

121. F. Alfert, "Cost comparison of dust explosion protection techniques available on the market," in Proceedings of the 2nd World Seminar on the Explosion Phenomenon and on the Application of Explosion Protection Techniques in Practice, EuropEx, Gent, Belgium, March 1996.

122. H. Janssens, "Sicherheit zu einem erschwinglichen Preis!," VDI-Berichte, no. 1601, pp. 271–279, 2001.

Chapter 2

Experimental Study of Constant Volume Sulfur Dust Explosions

Joseph Kalman[1,2], Nick G. Glumac[1], and Herman Krier[1]

[1]University of Illinois at Urbana-Champaign, 1206 W. Green Street, Urbana, IL 61801, USA

[2]Naval Air Warfare Center Weapons Division, 1 Administration Circle, China Lake, CA 93555, USA

ABSTRACT

Dust flames have been studied for decades because of their importance in industrial safety and accident prevention. Recently, dust flames have become a promising candidate to counter biological warfare. Sulfur in particular is one of the elements that is of interest, but sulfur dust flames are not well understood. Flame temperature and flame speed were measured for sulfur flames with particle concentrations of 280 and 560 g/m^3 and oxygen concentration between 10% and 42% by volume. The flame temperature increased with oxygen concentration from approximately 900 K for the 10% oxygen cases to temperatures exceeding 2000 K under oxygen enriched conditions. The temperature was also observed to increase slightly with particle concentration. The flame speed was observed to increase from approximately 10 cm/s with 10% oxygen to 57 and 81 cm/s with 42% oxygen for the 280 and 560 g/m^3 cases, respectively. A scaling analysis determined that flames burning in 21% and 42% oxygen are diffusion limited. Finally, it was determined that pressure-time data may likely be used to measure flame speed in constant volume dust explosions.

INTRODUCTION

Sulfur dust cloud combustion is a potential candidate to counter biological weapons. Sulfur dust has been used as a pesticide [1]. However, the more intriguing aspect of sulfur dust clouds is that they produce sulfur oxides which are chemical precursors to sulfuric acid [1]. It is well-known that sulfuric acid is extremely corrosive and dangerous to living organisms. The concept is that burning sulfur clouds will produce sulfur oxides. In the presence of water, sulfuric acid can be formed. It is thought that the sulfuric acid created, coupled with the elevated temperatures and ultraviolet radiation produced, will kill the spores. Other strong acids have been shown to have sporicidal tendencies [2].

In recent years, compositions have been studied for biodefeat applications. Mechanical alloys of aluminum and iodine [3, 4] as well as aluminum-iodine pentoxide thermites have been studied [5] and shown to be effective in killing biological spores, at least in part, due to the release of iodine. Other mechanical alloys of titanium and boron have also been investigated for this purpose [3, 6, 7].

Fundamentally, sulfur dust flames are unique. Sulfur is one of two elemental dusts whose combustion products are gaseous at standard conditions (298 K, 1 atm) with carbon being the other. Unlike carbon, the melting and boiling points of sulfur are much lower than the adiabatic flame temperature of a sulfur-air flame. This combination of factors provides conditions for a dust flame with a potentially strong gas-phase component.

To date, very few studies have investigated sulfur dust clouds. The limited work on sulfur dust flames by Proust [8] is among the only publications to do so; however, that work has provided few details into the combustion mechanism. Therefore, the primary goal for the current work is to measure fundamental aspects of sulfur dust cloud combustion in terms of fundamental quantities (e.g., flame speed) and to gain insight into the physical and chemical mechanisms involved. Moreover, the use of pressure-time data from constant volume dust explosions to determine flame speed is investigated.

EXPERIMENTAL METHODS

The current study uses a 31 L cube chamber to maximize optical access. Five of the sides (including the door) have acrylic windows with circular viewing areas of 6.7 in diameter. Each window is clamped onto the side of the chamber with a size 6 pipe flange. Gas, vacuum, and pressure transducer ports are located on the top and sides of the chamber. A piezoresistive Kulite pressure transducer (XTM-190-250G) is used, and the signals are conditioned by an Endevco PR Conditioner model 106. The bottom (sixth side) has five 1 in diameter ports. One port is placed in the center with the other four being 3 inches away from the center, each in a different direction (i.e., left, right, front, and back). A single off-center port is used for wire feedthrough for the ignition source.

The port in the center of the chamber has a nozzle with forty, 0.889mm diameter holes (number 65 drill bit) at a 45 degree angle. The two-piece particle injector is mounted underneath the center port of the chamber. The first piece is attached to the chamber through 1/4 in-20 screws. It contains a port on the side that attaches to a compressed air line and 1/4 in stainless steel tube which extends to the center and is bent 90 degrees downward. The powder is placed in an aluminum holder with a conical bottom (from a 1 inch drill bit) that attaches to the second piece of the injector. The centered 1/4 inch tube elbow directs the air burst downward into the powder holder. The pressurized burst rebounds off the bottom of the powder holder and carries the particles upwards through the nozzle into the chamber. The chamber is sealed by o-rings on the windows, injector, and door. Additional descriptions of the chamber are provided in [9].

Sulfur powder (−325 mesh) from Alfa Aesar was used. Particle size analysis was conducted using a Jeol 6060LV scanning electron microscope (SEM). The particles had an average diameter of 22.4 μm with a Sauter mean diameter of 30.4 μm. An anticaking agent, Aerosil 200 (Evonik), was used to improve the dispersion characteristics of sulfur. The average diameter of the Aerosil 200 powder was 12 nm according to the manufacturer. A size distribution was not measured as the particles were too small to be resolved by the SEM. The sulfur was mixed with 1% of Aerosil 200 by mass in a low energy tumbler for 3 hours. The addition of the anticaking agent was observed to have a noticeable effect on the flowability and dispersal of sulfur. Further details on the choice of anticaking agent are given in [9].

A known mass of the sulfur mixture was placed within the injector. The mass of powder remaining in the injector after the test was measured to determine the actual amount of powder injected. Alligator clips on the ignition posts held the igniter in the center of the chamber. A 4 J igniter (pyrogen covered bridgewire, Estes) was used to initiate combustion. The charge was ignited by discharging a 1 F capacitor at 4000 V from an RISI fireset (Model FS-43).

Prior to the test, the chamber was put under vacuum and filled to slightly below atmospheric pressure (2 in Hg) so that the gas used for injection brought the total pressure to 1 atmosphere. A 100 psig burst of air was used to inject the powder. The injection lasted a total of one second under constant pressure. Ignition occurred 400 ms after injection had ended to allow for a uniform cloud to form and turbulence to dissipate.

The determination of this ignition time was made by analysis of two-dimensional particle concentration measurements. These laser extinction measurements provided quantitative information on how the particle concentration developed in time and space. The pixel intensities from images taken prior to powder injection were averaged and used as the incident intensity, I_0, in Beers Law:

$$\frac{I}{I_0} = \exp\frac{-3Q_{ext}LC}{2\rho D_{32}}. \quad (1)$$

The intensity of each pixel in subsequent images allowed for the particle concentration, C, to be determined since the extinction efficiency (Q_{ext}) from Mie theory, Sauter mean diameter, D_{32}, path length, L (355 mm), and density of sulfur, ρ were all known. The arithmetic mean and standard deviation of the particle concentration for all pixels were calculated to provide a statistical analysis of the uniformity of the cloud. The ignition time was taken as the point in time where the spatially averaged concentration approached the expected value (based upon the amount of powder injected) and the standard deviation approached a minimum or decreased.

This method also provided visual evidence of the turbulence dissipating. Although the turbulence was never directly measured, it is believed that this quantitative measurement of the particle concentration yields sufficient insight into the injection process. Moreover, all of the experimental conditions in this work have the same injection and ignition conditions (e.g., ignition delay and injection back pressure). Comparison of the results from each conditions should see minimal effects from the turbulence. The reader is directed to [9] for additional details and discussion on these measurements.

Diagnostics

Flame speed measurements were made with ionization probes, similar to the work of Nair and Gupta [10], but were applied to dust flames instead of the gaseous flames that they studied. Two of the off-center 1 in diameter holes were used to feedthrough wires for the ionization probes where the wires were terminated by an R-type thermocouple connector (Omega). Two 0.01 in diameter tungsten wires were fed through a two-bore, 1/8 in outer diameter aluminum oxide tube. Approximately 1/4 inch of each wire extended outside of the ceramic tube. The other end of each wire was attached to the complementary thermocouple connector. The placement of the ionization probes are shown in Figure 1 where the relative location of the two ionization probe combs were chosen to provide information of the symmetry of the flame. The distance between the probes in each comb was 15.5 ± 0.76 mm.

Journal of Combustion

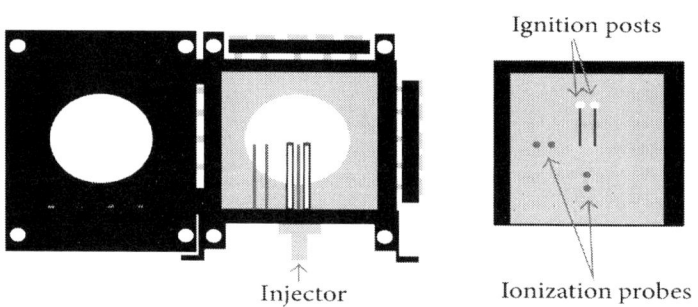

Figure 1: Schematic of the flame speed experimental setup.

Signals were recorded using a pair of Picoscope 4424 oscilloscopes, which collected 200,000 samples over a 2-second period. A pair of ionization probe traces from a sulfur flame is shown in Figure 3. The time when the first voltage drop reached a minimum for each trace was taken as the time when the flame front reached the probe. Laser shadowgraph confirmed this arrival qualitatively, as shown in Figure 2. The red curve indicates the location of the flame front. This feature was observed to reach the ionization probe at the same time (i.e., within the temporal resolution of the camera) the voltage trace spiked from the ionization probe (Figure 3). The shadowgraph measurement also provided qualitative information on the flame shape. Although portions of the front are smooth, the flame front is not symmetric. The cause of this appearance is likely a combination of remaining turbulence and natural convection.

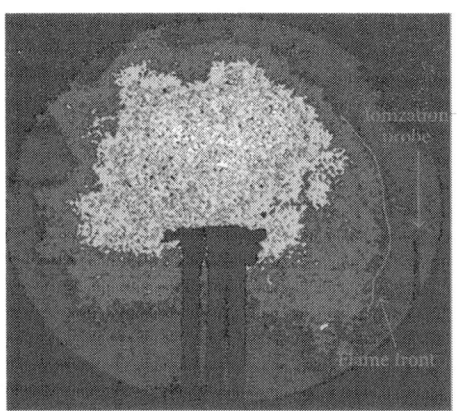

Figure 2: Qualitative shadowgraph measurement used to verify the time-of-arrival of the flame to the ionization probe. Image was taken prior to the flame reaching the probe.

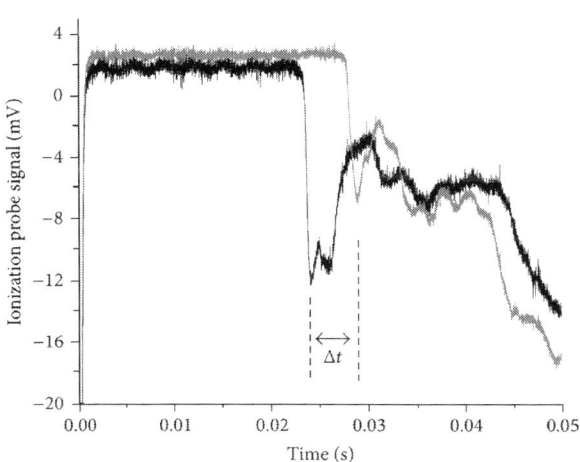

Figure 3: Representative voltage traces from the ionization probes from a sulfur flame.

The Δt shown in Figure 3 was used to calculate the flame speed. An uncertainty of 20% of the nominal value was associated with each measurement which is mostly due to the time-of-arrival measurement. xThe density corrections were calculated using Cantera [11] and the SOx mechanism from the University of Leeds [12].

Temperature measurements were made using thermocouples and pyrometry. Thermocouples (50 μm R-type, Omega part P13R-002) were covered in a thin coating of an aluminum oxide spray paint (ZYP Coating, A aerosol) to minimize catalytic effects. Thermocouples were attached to the connectors within the dust explosion chamber in the same manner as the previously described ionization probes. The thermocouple extension wire was connected to an Omega data acquasition (DAQ) system (Omega part OMB-DAQ-3005). Temperature was sampled every 100 μs for the first 2 seconds of each test. The DAQ system was triggered by a TTL pulse generated by the delay generator. The measurements were corrected for radiative and conductive losses.

Pyrometry measures the temperature of the condensed phases by comparing the thermal background to Planck's equation (with the emissivity included). A three-color pyrometer was used to obtain time-resolved temperature information by monitoring the emission at 700, 825, and 900 nm. Hamamatsu R928 photomultiplier tubes (PMT) were used for the 700 and 825 nm channels, while an R636-10 PMT was used for the 900 nm channel. Light was collected into a trifurcated fiber optic cable where each of the three branches went to a different PMT. A Stanford Research System (SRS) 300 MHz quad preamplifier (SRS model SR445) conditioned each signal before being recorded by the Picoscope.

A fiber optic-coupled Ocean Optics Jaz spectrometer was also used for pyrometry measurements. This spectrometer records spectra from 200 to 870 nm.However, due to spectral features from SO, SO_2, and S_2 from the ultraviolet into the visible region of the spectrum, only the thermal emission in the range of 600 to 850nm was used to determine the condensed phase temperature. The intensity calibration for both devices was conducted with a tungsten lamp (Ocean Optics LS Cal 1) with a known spectral intensity for the spectral regions studied.

RESULTS AND DISCUSSION

Measurements of pressure rise, temperature, and flame speed were made for 6 different conditions. Two different concentrations of the sulfur/anticaking agent mixture were used (280 and 560 g/m^3) and three different concentrations of oxygen (10, 21, and 42% by volume). The remaining balance of the gas was nitrogen. The stoichiometric conditions were 280 g/m^3 in 21% oxygen and 560 g/m^3 in 42% oxygen. The choice of these conditions was based upon consideration of the application (i.e., burning in air) and being able to isolate the effects of particle concentration and stoichiometry. Difficulties establishing flames at lower particle concentrations prevented the use of those concentrations. The tendency of sulfur to agglomerate, especially at the higher sulfur concentrations, yielded an upper limit on the particle concentrations used. The problem of agglomeration is addressed in greater detail in [9].

Pressure Data

A representative pressure-time curve and its first temporal derivative are shown in Figure 4. The maximum pressure rises are shown in Figure 5 with the theoretical maximum pressure rise and were determined from NASA's CEA program [13] under constant volume. It was observed that the pressure rise increased with particle concentration and increasing oxygen concentration. The data is consistent with work by Cashdollar [14] after scaling their pressure data to account for the different chamber volumes between the facilities used in each respective study.

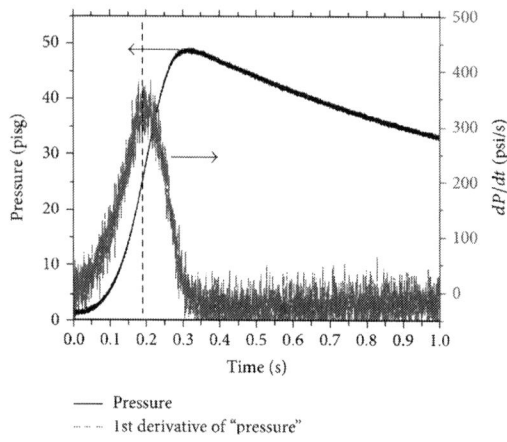

Figure 4: Sample pressure-time data from a sulfur explosion.

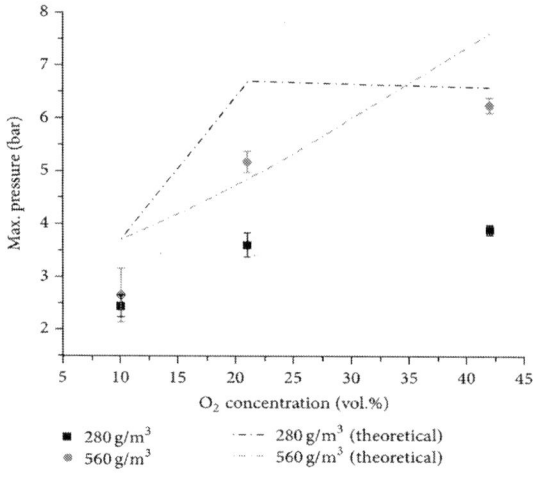

Figure 5: Maximum pressure rise (absolute pressure) for sulfur explosions within a 31 L chamber.

The 560 g/m³ condition within a 21% oxygen environment resulted in a pressure rise greater than the maximum pressure. This result may be due to particle settling from increased agglomerations during the experiment. An increase in agglomerations at the higher particle loadings were reported in [9]. If additional particles fall out of the suspension, it will bring the equivalence ratio closer to stoichiometric for these conditions. The maximum pressure will increase and approach the value seen for the lower particle loading. A similar effect of particle settling should then be expected for the 10% and 42% oxygen conditions. The difference in the maximum pressure rise for 10% O_2 is negligible because the maximum theoretical pressure rises are almost identical. Figure 5 shows that the maximum pressure (dashed lines) in the oxygen enriched case would be lower if the particle settling increased. This decrease is consistent with what was observed experimentally. For all of these conditions, it is challenging to quantify how much the pressure should change theoretically because the actual mass of settled particles is unknown.

The maximum rate of pressure rise, Figure 6, was seen to increase monotonically with oxygen concentration. For oxygen concentrations above 10% by volume, the rate of pressure rise is greater for higher particle concentration. The rate of pressure rise indicates how quickly the heat is being release. The importance of this quantity will be seen in (2), where the rate of maximum pressure rise is directly proportional to the flame speed.

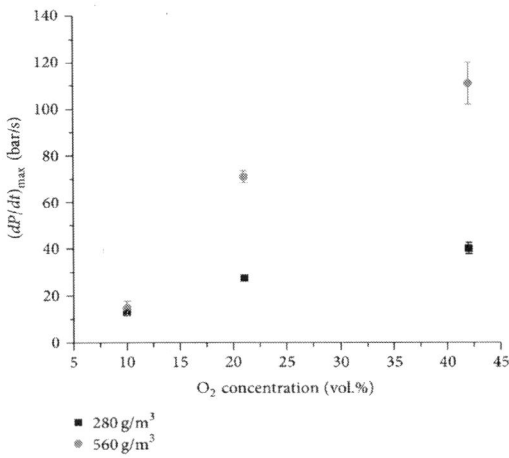

Figure 6: Maximum rate of pressure rise (absolute pressure) within the 31 L chamber from sulfur explosions.

Validity of Pressure-Time Data for Flame Speed Measurements

Pressure-time data may also be used to determine flame speed. Dahoe and de Goey [15] analyzed constant volume explosions (not necessarily dust explosions) from a thermodynamic standpoint to relate pressure-time data to laminar flame speed. This equation is shown as

$$\frac{dP}{dt} = \frac{3}{R}\left(\frac{dx}{dP}\right)^{-1}\left[1 - \left(\frac{P_i}{P}\right)^{1/\gamma}(1-x)\right]^{2/3}\left(\frac{P}{P_i}\right)^{1/\gamma} S_L, \qquad (2)$$

Where

$$x(P) = \frac{P - P_i}{P_e - P_i} \qquad (3)$$

and P is the instantaneous pressure, P_i and P_e are the initial and maximum pressures, respectively, and R is the spherical equivalent radius taken to be the radius of sphere with the same volume as the chamber used in the current study. The ratio of the specific heats, γ, was assumed to be constant (and approximately equal to 1.4 here because of the use of air), and S_L is the laminar flame speed. The analysis Dahoe and de Goey [15] conducted on this method included the additional assumption of a linear dependence on pressure for the mass burnt fraction, $x(P)$ (3). The work by Luijten et al. [16] used a multizone approach to develop a more rigorous definition for x, although it is not shown here because of its length.

The derivation of (2) was based on ideal assumptions that in practice may not be true for dust explosions. They are as follows.
- The chamber is well-insulated and assumed to be filled with reactants that are perfectly mixed and stagnant.

- The mixture is ignited from the center and the flame front produced is spherical, infinitely thin, and not wrinkled.
- The flame front breaks the chamber into two regions, the burnt and unburnt gases, where the mixtures within each zone are uniform (e.g., composition and temperature).
- Since the chamber has a fixed volume and heat cannot escape due to the well-insulated walls, the pressure rises from the heat addition. Moreover, the hot combustion products within the spherical flame will expand, thus compressing the unburnt gas isentropically.

Two concentrations of the sulfur mixture were tested to determine if the use of pressure-time data to measure flame speed is appropriate for dust explosions. A near stoichiometric concentration of 264 ± 34 g/m3 and a fuel-rich condition of 498 ± 53 g/m3 were ignited in air by a pyrotechnic igniter. Energy release by the igniter did not have a significant influence on the measurement [9]. The uncertainty in the concentration is due to the distribution of the measured powder mass injected.

The two combs of ionization probes discussed above were used to directly measure the flame propagation speed and compared to those calculated from (2) and (3). The calculated flame speed based upon the pressure-time data shown in Figure 4 is displayed in Figure 7.

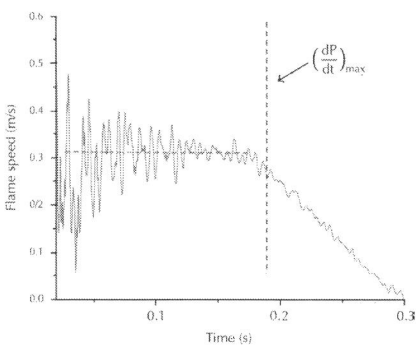

Figure 7: Calculate flame speed from pressure-time data.

The calculated laminar flame speed initially oscillates significantly due to a low signal-to-noise ratio from the pressure transducer signal. These oscillations dampen as the pressure rises. Despite the oscillatory nature, each calculated curve was observed to oscillate around a constant value until the time where dP/dt is a maximum. The time of the maximum rate of pressure rise is indicated by the vertical lines in Figures 4 and 7. The laminar flame speed from the pressure data was taken as the average speed from 20 ms after ignition to the time where $(dP/dt)_{max}$ was reached. The time of 20ms was used to limit the influence of the low signal-to-noise ratio of the pressure data very close to the instant of ignition. The decrease in laminar flame speed after $(dP/dt)_{max}$ was also observed by Santhanam et al. [17]. The reason for the decrease in flame speed is very likely due to heat losses. As the flame approaches the walls of the vessel, additional heat is lost to those walls which are near room temperature. The rate of pressure rise, which is related to the rate of heat release, decreases because some of that energy is absorbed by the chamber walls. The rate of pressure rise is proportional to flame speed (see (2)) so that a decrease in flame speed is observed. An analysis on the primary mode of heat transfer from the flame is discussed in the following section.

Heat Loss Analysis

The decrease in flame speed after $(dP/dt)_{max}$ was believed to be due to heat losses. With the increased amount of thermal radiation from dust flames, it is necessary to determine the importance of the various modes of heat transfer, specifically conduction and radiation.

The amount of energy lost to the chamber walls is difficult to quantify because of the complexity of the problem. However, the relative importance of conduction versus radiative losses may be analyzed qualitatively. The ratio of conductive to radiative losses, (4), was approximated for different flame temperatures and location (i.e., distance between the flame and the wall) where k is the thermal conductivity, ϵ is the emissivity, σ is the Stefan-Boltzmann constant, x is the distance between the flame and the wall, and T is the temperature at the location indicated by its subscript. The variable F is the ratio of the surface area of the particles within the flame and flame front surface area. Since the emissivity of the powder is unknown, the ratio of the heat losses was calculated as a function of emissivity. Figure 8 shows the results from these calculations. Consider

$$\frac{Q_{cond}}{Q_{rad}} \approx \frac{k}{F\epsilon\sigma} \frac{(T_{flame} - T_{wall})}{\Delta L} \frac{1}{T_{flame}^4 - T_{wall}^4}. \tag{4}$$

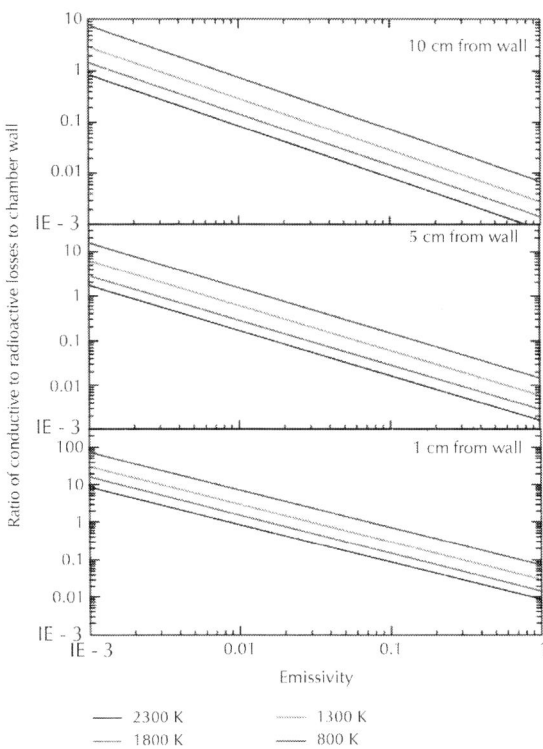

Figure 8: Calculated (first-order approximation) ratio of the conductive and radiative heat losses at multiple positions, at flame temperatures, and for F=1.

The calculated ratios displayed in Figure 8 assumed the ratio of the particle surface area to that of the flame was unity. This assumption is not realistic since the particles only constitute a fraction of the flame front for a given flame thickness. The value of F can be determined using geometric considerations, an average particle diameter, and a flame thickness. Santhanam et al. [17] calculated the flame thickness of Al dust flames by multiplying the flame speed by the burntime of an individual Al particle. This same estimation is more challenging for sulfur dust since the burn time of an individual particle is unknown. The effect of flame thickness will be discussed.

High temperatures and large values of the emissivity favor radiative dominated heat losses when F=1, which is to be expected since thermal radiation is dependent on those two parameters. For most of the conditions in this calculation set, radiation is either dominant or comparable to conduction, as illustrated in Figure 8. However, when the actual surface area of the particles compared to the flame front is considered, the value of F decreases from unity to approximately 0.35 and 0.038 when the flame is 1 and 10 mm thick, respectively. This factor will drive the ratio from (4) towards conduction. Therefore, if the flame thickness is in fact on the order of several millimeters or larger, it is unlikely that observed decrease in calculated flame speed is dominated by radiative losses unless the emissivity, temperature, and flame diameter (i.e., close to the wall) are large. Although further information is needed to determine the flame thickness accurately, this finding is still significant. Heat losses from sulfur dust flames may not be dominated by radiation which is contrary to what has been concluded about other dust flames (e.g., Al) where the product is also solid [17].

Discussion

The measured flame speeds from all of the test for both measurement techniques are shown in Figure 9. A large amount of data scatter is seen for both concentrations. It is believed that the range of flame speeds observed is at least in part due to the turbulence that remained in the system after the injection process. The flame speeds measured in each direction for a single experiment were observed to vary more than 50% in some cases. Moreover, there were instances where the signals from two probes in a single pair would indicate the flame had arrived at the same time. This observation is likely due to the flame approaching the ionization probe pair from a side-on approach rather than head-on as would be expected for a spherical flame. This point is further supported by the asymmetry observed by the highly irregular front seen by shadowgraph measurements [9].

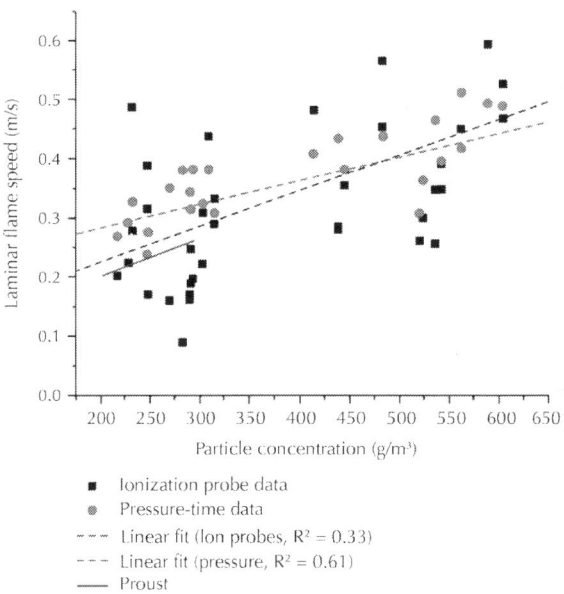

Figure 9: Comparison of the flame speed determined by the ionization probes, pressure-time data, and work from Proust [8].

The calculated laminar flame speeds from the pressure data and the ionization probes are plotted in Figure 9. Reasonable agreement is seen between the two measurement techniques. The flame speeds reported by Proust [8] were approximately 20% lower than the stoichiometric flame speed measured in the current work. This result suggests that it is plausible to use pressure-time data to estimate laminar flame speed. The higher flame speeds measured here are likely in part due to increased turbulence here, although it is difficult to prove without any doubt, because the level of turbulence was not measured in either study. Natural convection potentially also played a role. The burning sulfur particles produce hot gases, which of course will rise. The upward draft will then distort the flame front potentially increasing the surface area. Evidence of the effect of natural convection can be observed in Figure 2 where the flame has clearly propagated upwards more than downwards.

Therefore, it is somewhat inappropriate to use this technique to determine the laminar flame speed because of how both turbulence and natural convection effect the flame shape. It better represents the flame speed of an equivalent spherical flame with the same mass consumption rate. With that being said, the pressure-time data does provide a measure of a flame speed based upon the degree of turbulence. This result should still allow comparisons of flames in different conditions to be compared as long as the degree of turbulence is kept constant (i.e., injection parameters, ignition delay, and energy).

Flame Temperature

The temperature of the sulfur flames were measured by thermocouple and pyrometry. Both measurements were used to determine the peak temperature. The thermocouple provided the peak temperature locally, while pyrometry measurements indicated the peak temperature within the field of view of the pyrometer. The spatially integrated pyrometry signal is always biased towards the hottest regions due to the strong temperature dependence on the intensity of thermal radiation. Pyrometry data was collected with the 3-color PMT pyrometer and the Jaz Ocean Optics spectrometer. It should be noted that only the 825 nm and 905 nm signals were used because the 700 nm channel did not provide a sufficient signal level. The integration time for all data collected by the spectrometer was 10 ms.

Representative traces of the time-resolved data provided by the PMT pyrometer are displayed in Figure 10 with the calculated temperature using the gray body approximation. The early peak seen in Figure 10 was due to the emission from the pyrotechnic igniter. It was observed that the maximum temperature from the sulfur explosion occurred near the time that the rate of pressure rise was maximum. The peak temperature was recorded for each experiment.

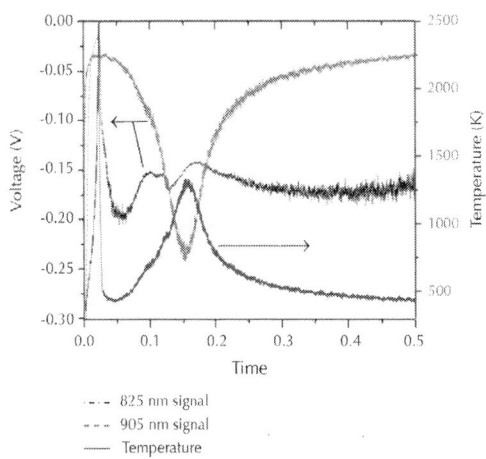

Figure 10: Representative filtered signals and calculated temperature for the PMT pyrometer.

Figure 11 shows a spectrum taken from the Jaz spectrometer. A gray body was believed to be the most appropriate assumption for the spectral emissivity because the λ^{-1} or λ^{-2} may not be appropriate for all materials [18] and the large optical depths produced by the dust cloud [19]. The measured peak temperatures from pyrometry and the thermocouples are shown in Figure 12. Only thermocouple data was obtained from the 10% oxygen tests, because the pyrometry signals were very weak and much lower than the noise level. The PMT pyrometer was used only for the 21% because of a spectral interference near 900 nm when the oxygen level was increased to 42% by volume. The thermal background from the Jaz spectrometer was fit for each of the tests for the oxygen-enriched conditions. As such, no time-resolved data were obtained for these conditions.

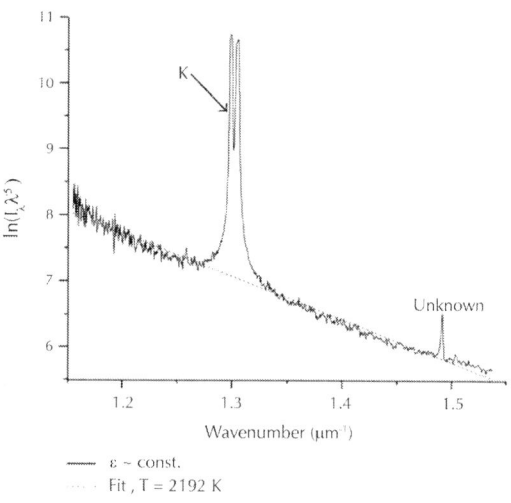

Figure 11: Fit of the thermal background from the Jaz spectrometer to determine temperature.

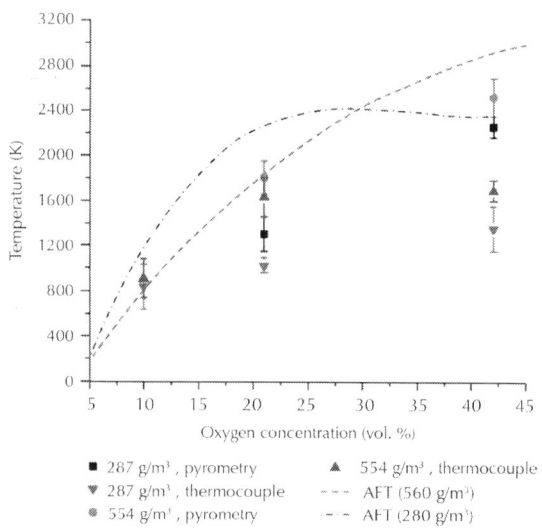

Figure 12: Temperature measurements of sulfur dust explosions.

The dashed lines represent the maximum flame temperatures calculated under equilibrium conditions. The temperature was adjusted by considering the heat absorbed by the anticaking agent. This change in temperature was minimal (typically less than 10 K). It should be noted that this adjustment is a first-order approximation since it did not include the equilibrium composition to be recalculated at the final temperature.

It was observed that the peak temperature steadily increased for the lower particle concentration from approximately 800 K to about 1300 K for the thermocouple measurements (corrected for radiative losses) as the oxygen content was increased from 10% to 42%. The increase in temperature from 21% to 42% oxygen from the pyrometry measurements was much greater as the maximum temperature exceeded 2000 K. This large difference is likely due to the fact that pyrometry measurements are biased towards the highest temperatures within the field-of-view while the thermocouples measured the local temperature.

A large temperature difference between the pyrometry and thermocouple measurements was also seen for the 21% and 42% oxygen tests at the higher particle loading. The temperature was observed to increase with particle loading for all oxygen concentrations. The average peak temperature for the 10% oxygen case was the lowest of the conditions for the 560 g/m^3 tests at 909 K.

The increase in temperature as oxygen concentration increased was likely due to the amount of energy liberated in each condition. The pressure data indicated that the maximum pressure rise (i.e., heat release), Figure 5, increased with oxygen concentration.

Flame Speed

Since the pressure-time data was determined to be an accurate way of measuring flame speed within the constant volume explosion chamber, it was decided to use that diagnostic to determine how particle and oxygen concentrations affected the flame speed. The same approach that was used to determine flame speed, as discussed above, was applied to the other conditions. The data for all of the particle and oxygen concentrations are summarized in Figure 13.

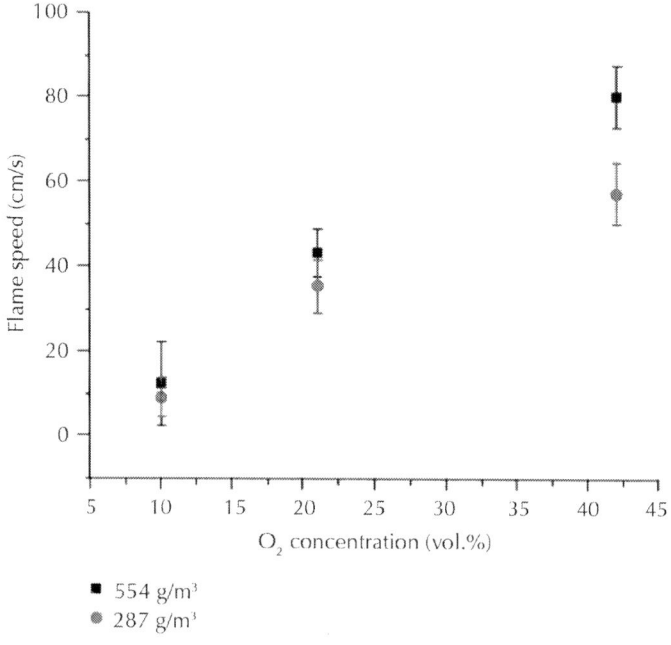

- ■ 554 g/m³
- ● 287 g/m³

Figure 13: Flame speed measurements by analysis of pressure-time data for sulfur dust explosions.

The individual data points for the 21% oxygen case were displayed in the previous section (see Figure 9). The averages at that oxygen concentration also includes additional data points taken after the completion of that portion of the work. A monotonic increase in flame speed was observed with an increase in oxygen concentration. A similar trend is shown in Figure 6 for the $(dP/dt)_{max}$ data. Since the flame speed was calculated from the pressure data (and S_L scales with dP/dt), it is expected to see a similar dependence. For a given oxygen concentration, the higher particle loading condition (i.e., higher equivalence ratio) has a greater flame speed.

COMBUSTION MECHANISM DISCUSSION

Gas-phase emission from sulfur oxides and S_2, as reported in [9], suggests that sulfur has a gas-phase component, but the extent of the gas-phase chemistry is unknown. However, the data do not explicitly suggest anything about the limiting process(es) during combustion, namely, if sulfur burns in a kinetically or diffusion limited manner.

Damköhler Number Analysis

Calculation of the key Damköhler number will aid in this discussion. This Damköhler number is the ratio of the chemical and diffusion time scales. A chemical time scale was determined by simulating an adiabatic, constant volume, perfectly-stirred reactor (PSR) with the SO_x mechanism from University of Leeds [12]. The ratios of sulfur, oxygen, and nitrogen that were present in the chamber after injection under atmospheric pressure were used as the initial conditions. An initial temperature of 1300 K was needed to start the reaction. The time step was set to 1 microsecond. The chemical time scales were taken as the time it takes for the temperature to go from 10% to 90% of the total temperature change. The temperature change was the difference from 1300 K to the adiabatic flame temperature.

The diffusion time scale is inversely proportional to the diffusion coefficient, D, which was calculated from kinetic theory [20]. The binary diffusion coefficient of oxygen through nitrogen was calculated using the measured peak temperatures. The diffusion coefficient scales as $T^{1.65}$ and $MW^{-0.5}$. Because of this scaling, even if the temperature were to increase by 1700K (bringing the temperature up to about 3100K, the adiabatic flame temperature, for the hottest measured condition), the diffusion time scale would only decrease by about a factor of about 2.8. Similarly, if the diffusion of two other species besides O_2 and N_2 were considered (i.e., SO, SO_2, and S_2), the diffusion time scale would change by less than a factor of 2 (increase or decrease). Therefore, the diffusion coefficient will not change significantly (i.e., an order of magnitude) for mixtures of sulfur compounds, oxygen, and nitrogen. The pressure was taken as 1 atm for this calculation. The reasoning for this assumption was that at the time when the flame front passed the thermocouples and ionizationprobes, the pressurewithin the chamber had not significantly risen.

The Damköhler numbers calculated are relative to one another (i.e., not absolute) to eliminate concerns on the appropriate area to use to nondimensionalize the value and since the maximum pressure does not change substantially (i.e., less than an order of magnitude). The Da numbers shown in Table 1 are normalized to the value at 21%O_2 at each respective particle concentration so that individual values have no meaning, but the relative change is important.

Table 1: Summary of the normalized Damköhler numbers

	10%O_2	21%O_2	42%O_2
280 g/m³	0.12	1.00	0.48
560 g/m³	0.80	1.00	7.00

A significant increase (i.e., by an order of magnitude) in Da is seen as the oxygen concentration increased from 10% to 21% for the lower particle concentration. A similar increase is observed for the 560 g/m³ concentration as the oxygen content rises from 21% to 42%. This increase suggests that as the oxygen concentration increases, diffusion becomes more important which at first glance seems counterintuitive. Considering that the oxygen was not the only quantity that changed, this result is justified. The temperature also increased and that affects the chemistry more significantly than diffusion (i.e., exponentially versus $T^{1.65}$). The exponent on temperature (1.65) includes the temperature dependence from the collision integral [21]. The more intriguing aspect of this observation is that for both of the particle loadings, diffusion becomes more important as the flame transitions from fuel-rich conditions to stoichiometric. Stoichiometry for the 280 g/m³ and 560 g/m³ concentrations is when there is 21% and 42% oxygen by volume, respectively.

Flame Speed Scaling Analysis

The above discussion only provided insight into how the flames in each condition burned relative to one another. Analysis of the flame speed and temperature can be used to further specify the combustion mechanism and any limiting phenomena.

The scaling of the flame speed should be dependent on the burning mechanism. Landau and Lifshitz [22] state that the flame speed for a thermally driven combustion wave scales as $(\alpha/\tau_{comb})^{0.5}$, where τ_{comb} is a combustion time scale. Goroshin et al. [23] argued that for a diffusion limited flame, τcomb scales inversely proportional to the diffusion coefficient of the gas. If the flame is diffusion limited, the flame speed should scale as $(\alpha D)^{0.5}$. Goroshin et al. [23] contended that if this thermally driven flame is kinetically limited, the difference in mass diffusivity of the gas mixtures (i.e., in each condition) should not play a role, thus the flame speed should scale with $(\alpha)^{0.5}$. This logic was used for the current work. The theoretical scaling for the diffusion and kinetically limited flames are shown in Tables 2 and 3, respectively. The velocity ratios from the data of the current work is displayed in Table 4.

Table 2: Flame speed scaling in the diffusion limit

	$V_{10}\%/V_{21}\%$	$V_{21}\%/V_{21}\%$	$V_{42}\%/V_{21}\%$
280 g/m³	0.64	1.00	1.73
560 g/m³	0.52	1.00	1.43

Table 3: Flame speed scaling in the kinetic limit

	$V_{10}\%/V_{21}\%$	$V_{21}\%/V_{21}\%$	$V_{42}\%/V_{21}\%$
280 g/m³	0.89	1.00	1.15
560 g/m³	0.85	1.00	1.09

Table 4: Measured flame speed ratios

	$V_{10}\%/V_{21}\%$	$V_{21}\%/V_{21}\%$	$V_{42}\%/V_{21}\%$
280 g/m³	0.25	1.00	1.62
560 g/m³	0.29	1.00	1.86

It is observed that the ratio of experimentally measured velocities from the 21% and 42% oxygen cases (for both particle concentrations) are much closer to the ratio predicted by the diffusion limited theory. This result, in conjuction with the previous discussion, suggests that oxygen enriched sulfur dust flames burn in the diffusion limit.

Intuitively, it would be expected that an oxygen enriched flame would burn in the kinetic limit since a higher concentration of oxygen is closer to the particle surface and potentially significantly reducing the diffusion time scale. However, the oxygen concentration is only one aspect of these flames. The temperature measurements show that the flame burns hotter as more oxygen was added to the system. The diffusion time scale does decrease with temperature ($T^{1.65}$), but it is not affected as much as the kinetics, which scale exponentially with temperature. The increase in temperature causes the chemistry to occur much faster, resulting in the diffusion process to be the limiting step.

Decreasing the oxygen concentration to 10% does not fit the scaling ratios predicted by kinetically or diffusion limited flame. The experiments with higher oxygen concentrations produced very consistent data whereas the 10% oxygen concentrations had a larger spread. The larger spread is not well represented by the data shown here. On multiple occasions, the 10% oxygen tests did not ignite (for both powder concentrations). It is possible that perhaps a flammability unit was approached by these oxygen depleted conditons.

Group Combustion Regime

Finally one must determine if the dust particles burn independently of each other. So a second condition that must be determined is the droplet spacing. The droplet combustion analysis assumes that there are no interactions with other particles [24]. The spray combustion community uses a group combustion number to estimate if the particles burn individually or together within a larger group flame. A group combustion number is defined by Glassman and Yetter [25] as follows:

$$G = 3\left(1 + 0.276 \text{Re}^{0.5} \text{Sc}^{0.5} \text{LeN}^{2/3}\right) \frac{R}{S}, \tag{5}$$

where Sc is the Schmidt number (ratio of momentum and mass diffusivities), Le is the Lewis number (ratio of thermal and mass diffusivities), N is the number of particles, R is the particle radius, and S is the average particle spacing [25]. The ratio of R and S is equal to the cube root of the quotient of the particle mass loading density (i.e., g/m³) and the particle density [26] and is on the order of 0.1 for the conditions tested in this work. The number of particles within the chamber for a given test is off the order of 10^{-8}-10^{-9}. Both Sc and Le will be on the order of 0.1 to 1. The Reynolds number is unknown because the velocity was not measured but it is likely orders of magnitude larger than 10^{-8}-10^{-9}. Therefore, using (5), will bemuch greater than 10^{-2}. A group number of less than 10^{-2} is specified for individually burning particles to occur [25].

CONCLUSIONS

Constant volume sulfur dust explosions were investigated. Measurements of flame speed using ionization probes showed reasonable agreement to the calculated speed from the pressure time data. Although there was agreement, it is inappropriate to call the quantity laminar flame speed because of the turbulence in the system. Flame speed was observed to range from 9 cm/s with 10% oxygen for both particle concetrations studied to as high as 80 cm/s in 42% oxygen for 554 g/m^3 of sulfur. The flame speed also increased with particle concentration which may be attributed to a shorter interparticle distance at higher concentrations. The temperature was measured to vary from approximately 800 K for both particle concentrations to approximately 2200 and 2600 for 287 g/m^3 and 554 g/m^3, respectively. Flame speed and temperature were not observed to be a function solely on equivalence ratio but rather dependent on the concentrations of sulfur and oxygen. Further analysis concluded that diffusion became the limiting process as the stoichiometry transitioned from fuel-rich to stoichiometric.

ACKNOWLEDGMENTS

This work was funded by DTRA Grant HDTRA1-11-1-0014 under project manager Dr. Suhithi Peiris. The authors would like to thank undergraduate students Sasank Vemulapati and Chris Murzyn for their assistance. This work was carried out in part in the Frederick Seitz Materials Research Laboratory Central Facilities at the University of Illinois at Urbana, Champaign.

REFERENCES

1. J. O. Nriagu, Ed., Sulfur in the Environment, Part I: The Atmospheric Cycle, John Wiley and Sons, 1978.
2. B. Setlow, C. A. Loshon, P. C. Genest, A. E. Cowan, C. Setlow, and P. Setlow, "Mechanisms of killing spores of Bacillus subtilis

by acid, alkali and ethanol," Journal of Applied Microbiology, vol. 92, no. 2, pp. 362–375, 2002.

3. D. Allen, Optical combustion measurements of novel energetic material in a heterogeneous shock tube [M.S. thesis], University of Illinois, Mechanical Science and Engineering, Champaign, Ill, USA, 2012.

4. S. Zhang, M. Schoenitz, and E. L. Dreizin, "Iodine release, oxidation, and ignition of mechanically alloyed Al-I composites," The Journal of Physical Chemistry C, vol. 114, no. 46, pp. 19653–19659, 2010.

5. B. R. Clark and M. L. Pantoya, "The aluminium and iodine pentoxide reaction for the destruction of spore forming bacteria," Physical Chemistry Chemical Physics, vol. 12, no. 39, pp. 12653–12657, 2010.

6. S. A. Grinshpun, A. Adhikari, M. Yermakov et al., "Inactivation of aerosolized Bacillus atrophaeus (BG) endospores and MS2 viruses by combustion of reactive materials," Environmental Science & Technology, vol. 46, no. 13, pp. 7334–7341, 2012.

7. M. Clemenson, Explosive initiation of various forms of the ti/2b energetic system [M.S. thesis], University of Illinois at Urbana-Champaign, Mechanical Science and Engineering, 2012.

8. C. Proust, "Flame propagation and combustion in some dust-air mixtures," Journal of Loss Prevention in the Process Industries, vol. 19, no. 1, pp. 89–100, 2006.

9. J. Kalman, Experimental investigation of constant volume sulfur dust explosions [Ph.D. thesis], University of Illinois at Urbana-Champaign, 2014.

10. M. R. S. Nair and M. C. Gupta, "Burning velocity measurement by bomb method," Combustion and Flame, vol. 22, no. 2, pp. 219–221, 1974.

11. D. Goodwin, Cantera: An Object-Oriented Software Toolkit for Chemical Kinetics, Thermodynamics, and Transport Processes, Caltech, Pasadena, Calif, USA, 2009.

12. T. Ziehn and A. S. Tomlin, "A global sensitivity study of sulfur chemistry in a premixed methane flame model using HDMR," International Journal of Chemical Kinetics, vol. 40, no. 11, pp. 742–753, 2008.
13. S. Gordon and B. J. McBride, "Computer program for calculation of complex chemical equilibrium compositions and applications," NASA Reference Publication 1311, NASA, Washington, DC, USA, 1996.
14. K. L. Cashdollar, "Flammability of metals and other elemental dust clouds," Process Safety Progress, vol. 13, no. 3, pp. 139–145, 1994.
15. A. E. Dahoe and L. P. H. de Goey, "On the determination of the laminar burning velocity from closed vessel gas explosions," Journal of Loss Prevention in the Process Industries, vol. 16, no. 6, pp. 457–478, 2003.
16. C. C. M. Luijten, E. Doosje, and L. P. H. de Goey, "Accurate analytical models for fractional pressure rise in constant volume combustion," International Journal of Thermal Sciences, vol. 48, no. 6, pp. 1213–1222, 2009.
17. P. R. Santhanam, V. K. Hoffmann, M. A. Trunov, and E. L. Dreizin, "Characteristics of aluminum combustion obtained from constant-volume explosion experiments," Combustion Science and Technology, vol. 182, no. 7, pp. 904–921, 2010.
18. J. Kalman, N. Glumac, and H. Krier, "High temperature metal oxide spectral emissivities for pyrometry applications," AIAA Journal of Thermophysics and Heat Transfer. Submitted.
19. J. Kalman, D. Allen, N. Glumac, and H. Krier, "Optical depth effects on aluminum oxide spectral emissivity," Journal of Thermophysics and Heat Transfer, vol. 29, no. 1, pp. 74–82, 2015.
20. N. M. Laurendeau, Statistical Thermodynamics: Fundamentals and Applications, Cambridge University Press, New York, NY, USA, 2010.

21. L. Monchick and E. A. Mason, "Transport properties of polar gases," The Journal of Chemical Physics, vol. 35, no. 5, pp. 1676–1697, 1961.
22. L. Landau and E. Lifshitz, Fluid Mechanics, Pergamon Press, 1959. View at MathSciNet
23. S. Goroshin, F.-D. Tang, A. J. Higgins, and J. H. Lee, "Laminar dust flames in a reduced-gravity environment," Acta Astronautica, vol. 68, no. 7, pp. 656–666, 2011.
24. S. R. Turns, An Introduction to Combustion: Concepts and Applications, McGraw-Hill Series in Mechanical Engineering, McGraw-Hill Education, 2000, http://books.google.com.eg/books?id=sqVIPgAACAAJ&redir_esc=y.
25. I. Glassman and R. Yetter, Combustion, Academic Press, 2008.
26. R. K. Eckhoff, "Chapter 4—propagation of flames in dust clouds," in Dust Explosions in the Process Industries, pp. 251–384, Gulf Professional Publishing, Burlington, Vt, USA, 3rd edition, 2003.

Chapter 3

Industry Specific Dust Explosion Likelihood Assessment Model with Case Studies

Junaid Hassan, Faisal Khan, Paul Amyotte, and Refaul Ferdous

Safety and Risk Engineering Group (SREG), Faculty of Engineering & Applied Science, Memorial University, St. Johns, NL, Canada A1B 3X5

Safety and Risk Engineering Group (SREG), Faculty of Engineering & Applied Science, Memorial University, St. Johns, NL, Canada A1B 3X5

Department of Process Engineering and Applied Science, Dalhousie University, Halifax, NS, Canada B3J 2X4

Safety and Risk Engineering Group (SREG), Faculty of Engineering & Applied Science, Memorial University, St. Johns, NL, Canada A1B 3X5

ABSTRACT

Dust explosion is a potential threat to the process facilities handling dusts. Dust explosion occurrences are frequently reported in these industries. Industrial professionals and researchers have been trying to develop effective measures to assess and mitigate and/or prevent dust explosion. To develop effective prevention and mitigation strategies, it is important to understand the interaction of dust explosion controlling parameters and also to assess likelihood of occurrence in given conditions. Authors have proposed a conceptual framework to model dust explosion likelihood. In this paper, a detailed implementation of the conceptual model is presented. Three different dust classes (i.e. food feed; plastic, resin and rubber; and metal alloys) are considered for model development. The proposed model considers six key parameters of dust explosion: dust particles diameter, minimum ignition energy, and minimum explosible concentration, minimum ignition temperature, limiting oxygen concentration and explosion pressure. These parameters are conditional to the type of dust and chemical composition. A conditional probabilistic approach is used to determine the total probability of dust explosion in a given process facility. Use of this model will help to assess the likelihood of dust explosion in given operating conditions. Moreover, it will help to develop prevention strategies focusing on the parameters that are responsible for dust explosion Three case studies are presented here to demonstrate the application of the model in real life.

INTRODUCTION

A dust explosion can take place when the suspended solid particles accumulated in the air receive sufficient energy from the source. The consequence is akin to a typical gas explosion in terms of the impact on the surrounding environment, industrial assets and monetary value. Unfortunately, the dust explosion's causation and severity are less familiar compared to the gas explosion among industrial practitioners.[1] for gas explosion, fuel, oxidant and ignition sources are necessary, while dust explosion requires two more vital criteria: appropriate mixing and confinement. These five elements are denoted with the dust explosion pentagon. The phase of the fuel during gas and dust explosion is different. Gas particles are in a gaseous phase, whereas dust particles are in a solid phase. Therefore, particle size of the dust is a very important fact on which to focus. According to the National Fire Protection Association (NFPA), any finely divided solid, 420 μm (micron) or 0.017 in. or less in diameter (i.e. material capable of passing through a U.S. No. 40 Standard sieve) is defined as dust.[2] The prime concern is combustible dust. Any dust capable of creating a violent explosion when it is suspended in air in ignitable concentrations, regardless of size, shape or chemical composition is called combustible dust.[1] The range of explosible particle size may be larger than the defined range for a specific material. Particle sizes distributions are often considered as a measure of the particle diameter in addition to the mean or median diameter.[1] In this paper, the median particle diameter is considered throughout the study.

A number of recent dust explosion phenomena caused severe loss to human lives and associated industries. On January 29, 2003, a massive dust explosion at the West Pharmaceutical Services facility in Kinston, North Carolina, killed six workers and destroyed the facility.[3] On February 20, 2003, a series of dust explosions at the CTA Acoustics facility in Corbin, Kentucky, killed seven workers, injured 37, and destroyed the facility.[3] On October 29, 2003, an aluminum dust fueled explosion killed one worker and injured several others at Hayes Lemmerz International in Huntington, Indiana.[4] On January 9, 2001, at the wool factory "Pettinatura Italiana" in Vigliano Biellese (BI), a massive explosion caused the death of three people, five severely injured personnel and considerable damage to part of the factory.[5] On February 7, 2008, a series of sugar dust explosions at the Imperial sugar manufacturing facility in Port Wentworth, Georgia, resulted in 14 worker fatalities.[6] With the increasing number of dust explosions in process facilities, the risk has become more alarming. However, substantial progress has been made through extensive research and development for better understanding of dust explosion dynamics. Preventing an ignition source and explosive dust clouds, explosion venting, automatic explosion suppression and good housekeeping are elaborately reported in existing literatures as the means of protective measures of dust explosions.[7]

Industry professionals and researchers are striving for more pragmatic and easily implementable solutions to prevent dust explosion phenomena. However, in the context of quantitative assessment, a predictive tool to assess the explosion probability in a particular industry is absent. In this paper, an effort has been made to establish a probabilistic model to assess dust explosion occurrence. The model is applied for three dust classes: food feed; plastic, resin and rubber; and metal alloys. Five parameters are identified as dust explosion influential parameters: particle diameter, minimum explosible concentration, minimum ignition energy, minimum ignition temperature and limiting oxygen concentration, whereas the maximum explosion pressure represents the severity of a dust explosion.

Five essential elements (e.g. fuel, oxidant, ignition source, mixing and confinement) form a dust explosion pentagon.[8] These five elements are represented by five influencing parameters of dust explosion. When these parameters reach the explosible range, dust explosion occurs.[9] Explosion may not occur if all parameters do not reach the explosible range.[10] A conceptual framework has earlier been developed by the authors which describes the method of assessing the dust explosion probability.[11] In this paper, the implementation of the earlier model is discussed elaborately for different dust classes.

Three case studies have been studied to demonstrate the applicability of the model. This paper attempts to use the existing information (experimental data) for a particular industry to develop the dust explosion assessment model. To assess the conditional probability, two parameters at a time have been considered to estimate the probability of explosion occurrence for a given industry. Estimating the conditional probability for each parameter and integrating them over a range provides the total probability of dust explosion occurrence. The model renders a nomograph as a quick assessment tool. For a particular industry, the model can assess the probability of explosion in the base condition (normal operating condition). Based on the assessment, the processing facility can implement safety measures (e.g. inherent safety, procedural safety, safety management system, etc.) and can develop effective prevention and mitigation strategies in the working environment.

METHODOLOGY FOR DUST EXPLOSION ASSESSMENT AND MATHEMATICAL MODELING

The proposed methodology to assess dust explosion likelihood is comprised of five steps as outlined in the conceptual model.[11] These steps are subdivided into several sub-steps. Figure 1 represents the framework of the proposed methodology. The main steps are given below; for details see the work on dust explosion likelihood assessment.[11]

- Hazard identification,
- Data collection,
- Data analysis,
- Probabilistic modeling,
- Nomograph development.

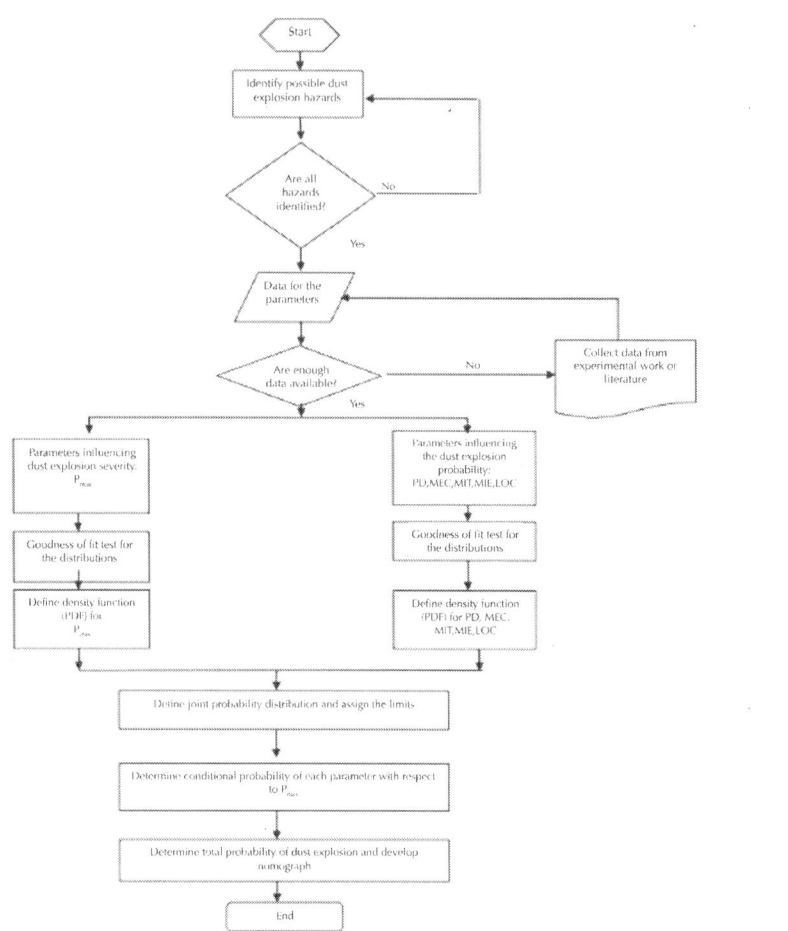

Figure 1: A framework for assessing dust explosion occurrence.[11]

MATHEMATICAL MODELING OF DUST EXPLOSION ASSESSMENT

The proposed methodology employs the rules of conditional probability. An elaborate description of the methodology is may be seen at Junaid et al.[11] Figure 1 highlight different steps of the methodology.

Model Testing

To use the model, probability distributions of the dust explosion parameters need to be determined for each dust class. The PDFs can be determined from the known distribution. These PDFs are used to formulate the joint probability distribution functions and are integrated over a range to get the CDFs. The integral range is identified according to the available data. Hence, the conditional probability values can be assessed for the particular dust classes. The total probability of dust explosion can be determined from the model and the nomograph is generated as a part of the model. The testing of the model is described in four steps as shown in Figure 1.

Data Collection

In this step, data for dust explosions parameters are collected. Six parameters are already identified in section "Methodology for dust explosion assessment and mathematical modeling". The data for analysis are collected from Echkoff's book (Appendix: Table A1 in reference[12]) for all the parameters except LOC. LOC data are excerpted from the NFPA.

Data Analysis and PDFs Determination

In this step, the collected data are analyzed to determine the underlying distributions of each identified parameter. This step provides the significant details of the distribution parameters. This information is used to determine the PDFs for each parameter. For example, consider a case where the statistical analysis for the hazard causing parameters is listed in Table 1 for a particular process facility.

Table 1: Dust Explosion Parameter Distribution Identification

Dust Explosion Parameter	Best Fitted Distribution	Estimated Distribution Parameter	95% Data Range
PD	Lognornal	$\lambda_{PD} = 4.02$, $\xi_{PD} = 0.955$	25–400 μm
MEC	Normal	$\mu_{MEC} = 80$, $\sigma_{MEC} = 45$	15–215 g/m³
MIT	Normal	$\mu_{MIT} = 504$, $\sigma_{MIT} = 65$	400–700 °C
MIE	Lognornal	$\lambda_{MIE} = 4.72$, $\xi_{MIE} = 1.5$	10–700 mJ
LOC	Normal	$\mu_{LOC} = 10.97$, $\sigma_{LOC} = 2.12$	8.75–12% O_2
P_{max}	Weibull	$\beta_{p}max=8.89$, $\theta_{Pmax}=10.7$	8–10.5 bar(g)

The estimated parameters of the identified distributions (from Table 1) can be used in the mathematical formulation of the probability distribution function. Based on the distribution types, formulated PDFs are given as:

For PD:

$$f_{PD}(PD, \lambda_{PD}, \xi_{PD}) = \frac{1}{PD \xi_{PD} \sqrt{2\pi}} e^{-(1/2)((\ln(PD)-\lambda_{PD})/(\xi_{PD}))^2} \quad (1)$$

For MEC:

$$f_{MEC}(MEC, \mu_{MEC}, \sigma_{MEC}) = \frac{1}{\sigma_{PD} \sqrt{2\pi}} e^{-(1/2)((MEC-\mu_{MEC})/\sigma_{MEC})^2} \quad (2)$$

For MIT:

$$f_{MIT}(MIT, \mu_{MIT}, \sigma_{MIT}) = \frac{1}{\sigma_{MIT} \sqrt{2\pi}} e^{-(1/2)((MIT-\mu_{MIT})/\sigma_{MIT})^2} \quad (3)$$

For MIE:

$$f_{MIE}(MIE, \lambda_{MIE}, \xi_{MIE})$$
$$= \frac{1}{MIE \xi_{MIE} \sqrt{2\pi}} e^{-(1/2)((\ln(MIE) - \lambda_{MIE})/\xi_{MIE})^2} \tag{4}$$

For LOC:

$$f_{LOC}(LOC, \mu_{LOC}, \sigma_{LOC})$$
$$= \frac{1}{\sigma_{LOC} \sqrt{2\pi}} e^{-(1/2)((LOC - \mu_{LOC})/\sigma_{LOC})^2} \tag{5}$$

For P_{max}:

$$f_{P_{max}}(P_{max}, \beta_{P_{max}}) = \frac{\beta_{P_{max}}}{\theta_{P_{max}}}$$

$$\left(\frac{P_{max}}{\theta_{P_{max}}}\right)^{\beta_{P_{max}} - 1} e^{-(P_{max}/\theta_{P_{max}})^{\beta_{P_{max}}}} \tag{6}$$

Identification of Integral Limits and CDFs Determination

The above Eqs. (1), (2), (3), (4), (5) and (7) are used in determining joint probability distribution functions and are integrated over a specific range to get the CDFs. The range of integral limits and CDFs are determined in this step.

For example, consider the parameter PD in the explanation. Data analysis for PD confirms that the range of 25–400 μm covers 95% of the data (Table 1) which is the addressed critical zone in this analysis. The limits for the integral reflect the vulnerable region where a dust explosion is likely. The vulnerable region is determined by assessing the data frequency in the specific range. The data used for the analysis are arranged and analyzed to understand the range. Data analysis for the severity parameter P_{max} confirms that most dust explosion pressure data are likely to be within 8–10.4 bar (g) 12. For the calculation, the upper bound is rounded up to 10.5 bar (g). Analyzing the dust explosion phenomenon in terms of P_{max} and PD alone, the probability of having a dust explosion due to the PD range susceptible to explosion may be written as Eq. (7):

$$P(P_{max}/PD) = \frac{\int_{P_{max}\text{critical}_{min}}^{P_{max}\text{critical}_{min}} \int_{PD\text{critical}_{min}}^{PD\text{critical}_{max}} P(P_{max} \cap PD) dPD dP_{max}}{\int_{PD\text{critical}_{min}}^{PD\text{critical}_{max}} PDP}$$

$$= \frac{\int_{25}^{400} \int_{8}^{10.5} P(P_{max} \cap PD) dPD dP_{max}}{\int_{1}^{420} PD dPD}$$

$$= \frac{\int_{25}^{400} \int_{8}^{10.5} C(f_{PD} \cdot f_{P_{max}}) * f_{PD}(PD; \lambda_{PD}, \zeta_{PD}) * f_{P_{max}}(P_{max}; \beta_{P_{max}}, \theta P_{max}) dPD dP_{max}}{\int_{1}^{420} f_{PD}(PD; \lambda_{PD}, \zeta_{PD}) dPD} \quad (7)$$

All the parameters and their values are already known and the functions for the PD and P_{max} are also determined from the distribution characteristics. Solving Eq. (7) provides the conditional probability of dust explosion due to the PD. The above process is repeated for the rest of the parameters (MEC, MIT, MIE and LOC) with respect to the severity parameter, P_{max}. The posterior probability density function is eventually transformed into the CDF to read the posterior conditional probability, according to the specific range as provided in Figure 2 for the above analysis.

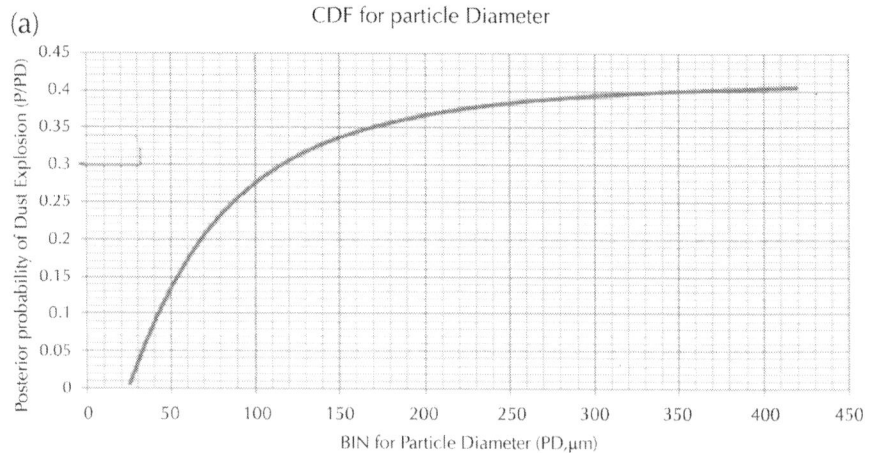

Industry Specific Dust Explosion Likelihood Assessment Model ...

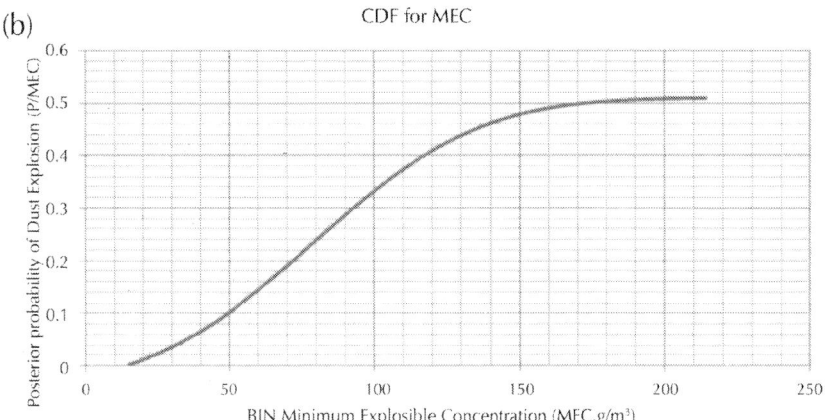

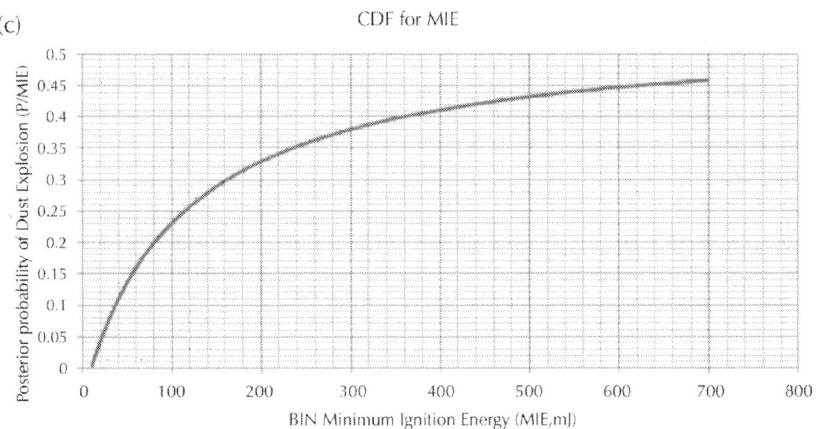

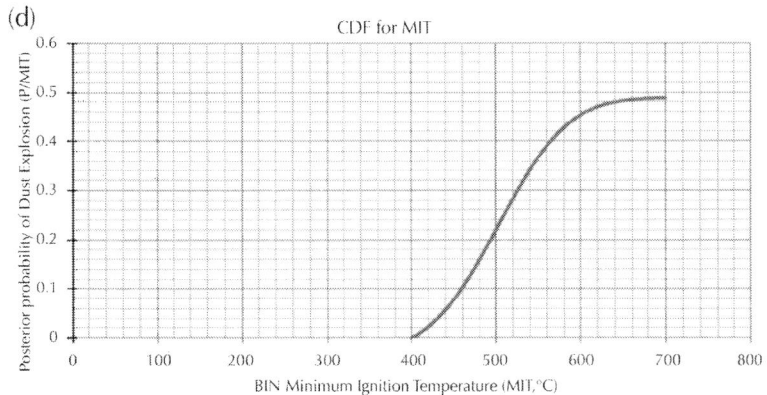

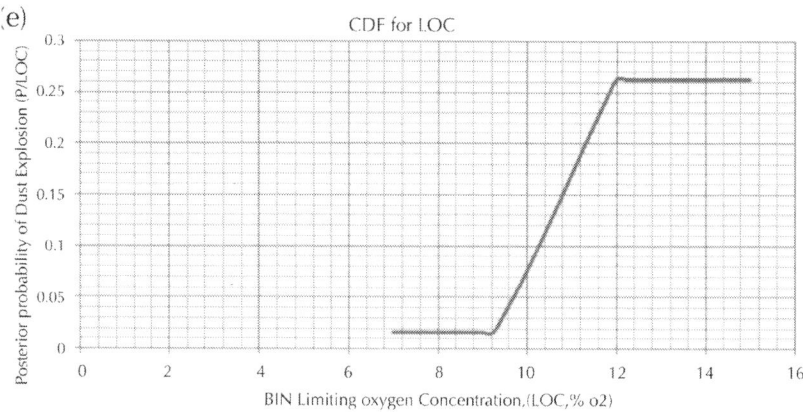

Figure 2: (a) Posterior probability for PD. (b) Posterior probability for MEC. (c) Posterior probability for MIE. (d) Posterior probability for MIT. (e) Posterior probability for LOC.

DUST EXPLOSION ASSESSMENT IN A GIVEN OPERATING CONDITION

The steps described in sections "Data collection, Data analysis and PDFs determination and Identification of integral limits and CDFs determination" develop the predictive model which can be used for a specific case in the considered process facility. To use the model, the operating conditions of the process facility are required. For instance, consider the following monitored data in the considered process facility. The following process operating condition is chosen to demonstrate the applicability of the model.

- PD varies from 40 μm to 350 μm
- MEC varies within 30 g/m³ to 200 g/m³
- Upper limit for MIT is 650 °C
- Upper limit of MIE is 600 mJ
- Upper limit for LOC is 11.75(% of O_2)

Industry Specific Dust Explosion Likelihood Assessment Model ...

For the given process parameters the predictive model analyses the data and provides the following result listed in Table 2. The predictive model also provides a nomograph as an easy way to interpret the result in a graphical form which is given in Figure 3.

Table 2: Nomograph Calculation Interface

Influencing Parameters	Lower Bound	Upper Bound	Conditional Probability
PD (25–400)	40	350	0.311 (P_{PD})
MEC (15–215)	30	200	0.473 (P_{MEC})
MIT (400–700)	400	650	0.482 (P_{MIT})
MIE (10–700)	10	600	0.444 (P_{MIE})
LOC (7–15)	7	11.75	0.223 (P_{LOC})
Total probability of dust explosion			0.151 (P_{Total})

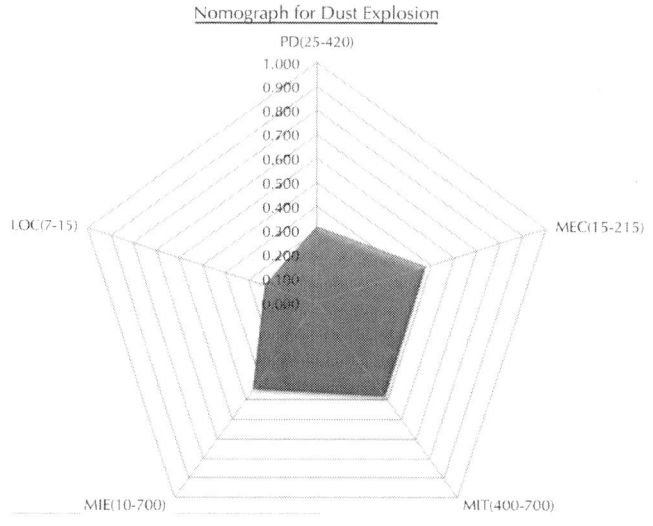

Figure 3: Nomograph for dust explosion.

The calculation depicted above indicates that the occurrence probability of dust explosion is 0.151 for the given conditions of the process facility. The nomograph indicates the probability of 0.151 inside the pentagon with the shaded portion (Figure 3). It shows that 1 out of 6 operations are susceptible to an explosion at the normal operating condition. The nomograph is a simple way to interpret the result of the complex mathematical equation in a graphic way. The process facility can formulate necessary prevention and mitigation strategies based on the assessment.

Identified Dust Classes and Case Studies

Three dust classes are identified to use the model in three specific industries. These classes are food feed; plastic, resin and rubber; and metal alloys. The industries dealing with the aforementioned dust classes can utilize the proposed model to assess the probability of dust explosion for a normal operating condition. The dust explosion assessment in a process facility may lead to significant modification of safety measures. It can be very helpful to formulate effective mitigation or prevention strategies. The proposed model includes three specific industries and they are discussed in the case studies below:

Dust Class 1: Food Feed

Dust produced or handled in the food processing industries is susceptible to explosion if necessary conditions are met. The food processing industries deal with dust such as dextrose, fructose, coffee, milk powder, wheat flour and sugar. Based on the statistical analysis on the available data of food feed, the dust explosion parameters and the distributions are provided in Table 3.

Table 3: Food Feed: Dust Explosion Parameter Distribution Identification

Dust Explosion Parameter	Best Fitted Distribution	Estimated Distribution Parameter	95% Data Range
PD	Lognormal	$\lambda_{PD} = 4.18, \xi_{PD} = 1.03$	23–400 μm
MEC	Lognormal	$\lambda_{MEC} = 4.18, \xi_{MEC} = 0.63$	31–700 g/m³
MIT	Lognormal	$\lambda_{MIT} = 6.19, \xi_{MIT} = 0.085$	441–580 °C
MIE	Lognormal	$\lambda_{MIE} = 5.29, \xi_{MIE} = 1.21$	76–1000 mJ
LOC	Normal	$\mu_{LOC} = 10.97, \sigma_{LOC} = 2.12$	8.75–12% O_2
P_{max}	Normal	$\mu_{P}max=7.93, \sigma_{Pmax}=1.43$	5.1–10.2 bar(g)

Taking the distribution parameter into account, the conditional PDFs are formulated. The analysis (in Table 3) also provides the integral range with which the conditional PDFs are integrated to obtain the CDFs. The CDFs are represented in Figure 4 as the CDF plot. The CDF plot represents the probability of dust explosion for different operating ranges. It facilitates the reading of the conditional probability of dust explosion for a given operating range of a single parameter.

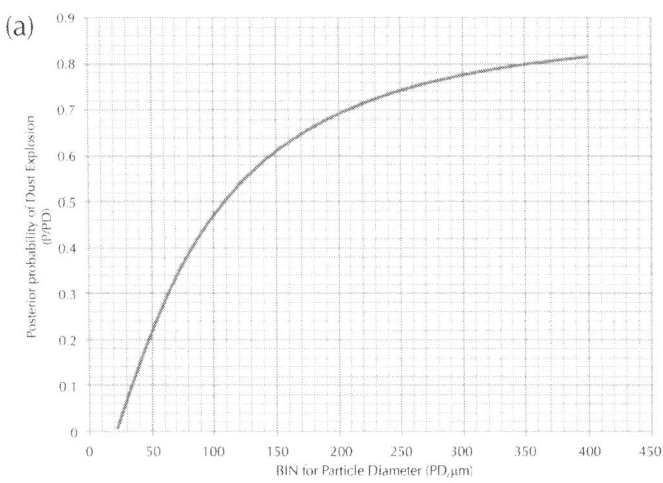

86 Dust Explosions in the Process Industries

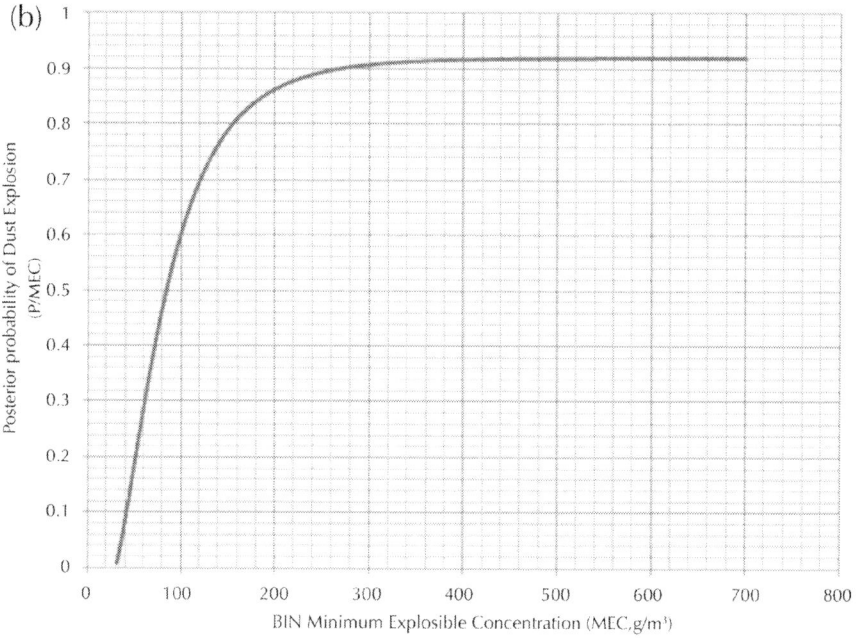

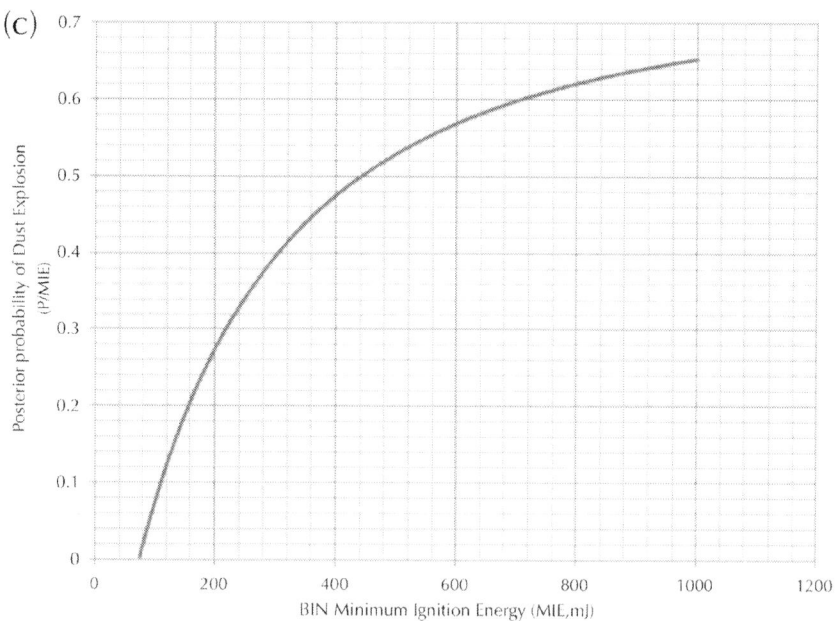

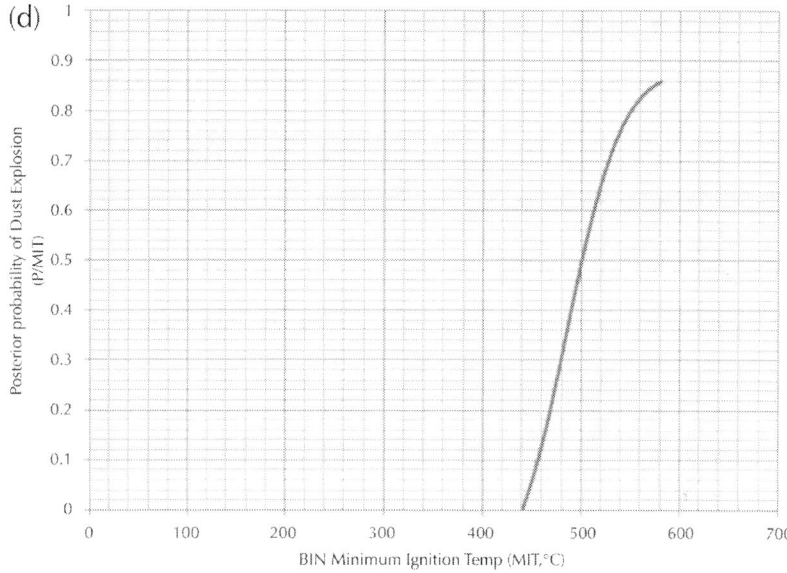

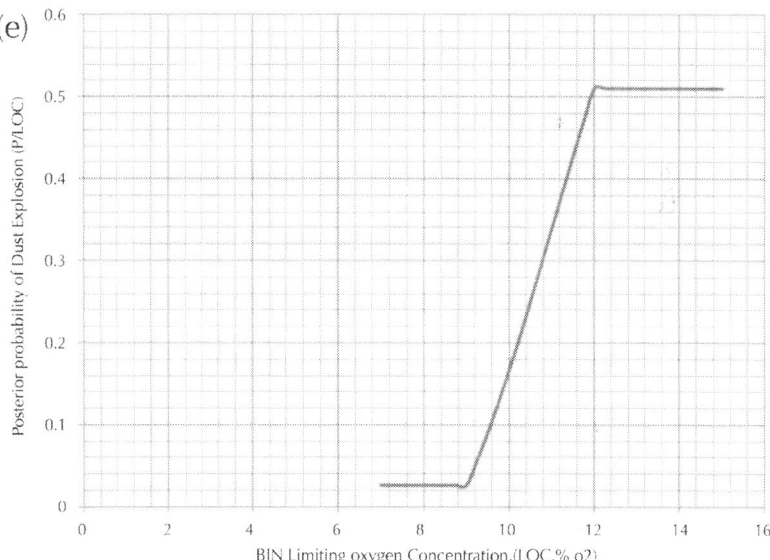

Figure 4: (a) Posterior probability for PD (dust class: food feed). (b) Posterior probability for MEC (dust class: food feed). (c) Posterior probability for MIE (dust class: food feed). (d) Posterior probability for MIT (dust class: food feed). (e) Posterior probability for LOC (dust class: food feed).

The total probability of dust explosion is assessed using the conditional probability of dust explosion for each parameter. The process of assessing the total probability of dust explosion based on the plant operating condition is described in section "Case study: sugar dust explosion and fire" with a case study.

Case Study: Sugar Dust Explosion and Fire

On February 7, 2008, at about 7:15 p.m., a series of sugar dust explosions took place at the Imperial Sugar manufacturing facility in Port Wentworth, Georgia. It resulted in fourteen workers fatalities and thirty-six workers were severely injured and burned which eventually caused permanent damage to them.[6] A sugar dust explosion occurred in the enclosed steel conveyor belt under the granulated sugar storage silos in the above facility. After a while, a massive secondary dust explosion propagated throughout the entire granulated and powdered sugar packing buildings, bulk sugar loading buildings, and parts of the raw sugar refinery.[6]

The proposed model for food feed considers granulated and powdered sugar, and an effort to assess the dust explosion probability at the Imperial sugar manufacturing facility has been made in this section. To assess the probability of dust explosion, understanding the operating conditions of the facility is required. This information is collected from relevant reports available on the facility and its operation. Any missing information regarding operational and material parameters is replaced with the proper engineering judgment. According to US CSB, the dust explosion parameters at the facility are reported as[6]:

- PD varies from 23 μm to 286 μm
- MEC range is 115 g/m^3
- Upper limit for MIT is 450–500 °C
- Upper limit of MIE is 1000 mJ
- Upper limit for LOC is 10.5(% of O$_2$)

The LOC and MIT values are chosen for the assessment. From the above plant condition the predictive model assesses the facility condition. The gist of the assessment is given in Table 4 and the model provides a nomograph (Figure 5) to understand the plant condition easily.

Table 4: Nomograph (Food Feed) Calculation Interface

Influencing Parameters	Lower Bound	Upper Bound	Conditional Probability
PD (23–400)	23	286	0.76 (P_{PD})
MEC (31–700)	31	115	0.68 (P_{MEC})
MIT (441–580)	450	500	0.44 (P_{MIT})
MIE (76–1000)	76	1000	0.65 (P_{MIE})
LOC (7.25–15)	7.25	10.50	0.19 (P_{LOC})
Total probability of dust explosion			0.27 (P_{Total})

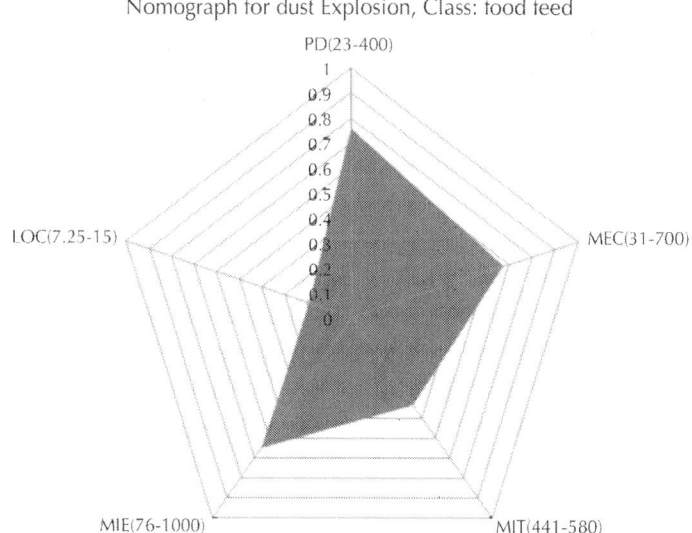

Figure 5: Nomograph (food feed) for dust explosion.

According to the proposed model, the total probability of dust explosion at the Imperial sugar manufacturing company is 27% when the base operating condition is considered. A dust explosion took place which indicates the probability reaches 100%, denoting the accident occurred. The accident means that something must have happened to trigger the base condition into becoming a catastrophic explosion. 27% probability of dust explosion is a high probability. It means one out of three facilities under normal operating condition is explosion prone. A target value of 0.001 or one in a thousand would be a reasonable value to choose. If the probability of dust explosion can be reduced in the normal operating condition, it will reduce the chance of explosion during process upset conditions as well. The model also interprets that the PD, MEC and MIE make significant contributions in the conditional probability (from Table 4). At the facility, the primary explosion took place at the enclosed steel belt conveyor where the dust concentration accumulated to the explosive range and an ignition source provided the necessary energy source to begin the explosion,[6] which validates the assessment of the model. To improve the safety of the facility, necessary design modification is needed to the steel belt conveyor and adequate housekeeping is required. These provide a better and safer condition for working. The model provides a quick estimate of the plant condition. Based on the assessment administrative controls, engineering design and operational modification can be done to reduce the total probability of dust explosion.

Dust Class 2: Plastic, Resin and Rubber

Industries dealing with the following dust: rubber, melamine resin, polyester, polyvinyl–alcohol, etc. are considered in this class. Based on the statistical analysis on the available data for afore-mentioned dust classes, the probability distribution type and details of the dust explosion parameters are listed in Table 5.

Table 5: Plastic, Resin and Rubber: Dust Explosion Parameter Distribution Identification

Dust Explosion Parameter	Best Fitted Distribution	Estimated Distribution Parameter	95% Data Range
PD	Lognormal	$\lambda_{PD} = 4.11, \xi_{PD} = 0.93$	21–280 µm
MEC	Lognormal	$\lambda_{MEC} = 3.99, \xi_{MEC} = 0.71$	31–200 g/m³
MIT	Normal	$\mu_{MIT} = 528, \sigma_{MIT} = 108.54$	421–790 °C
MIE	Lognornal	$\lambda_{MIE} = 5.18, \xi_{MIE} = 2.64$	14–2000 mJ
LOC	Normal	$\mu_{LOC} = 10.97, \sigma_{LOC} = 2.12$	8.75–12% O_2
P_{max}	Normal	$\mu_{p}max=8.432, \sigma_{Pmax}=1.04$	6.2–10.2 bar(g)

The model analyses the data and provides the conditional probability with respect to each parameter. The analysis is depicted with a CDF plot and represented by Figure 6. For a single parameter at any condition within operational range the plot can provide the conditional probability of dust explosion. The assessment of the total probability of dust explosion from the conditional probabilities is discussed in section "Case study: dust explosion in wool factory".

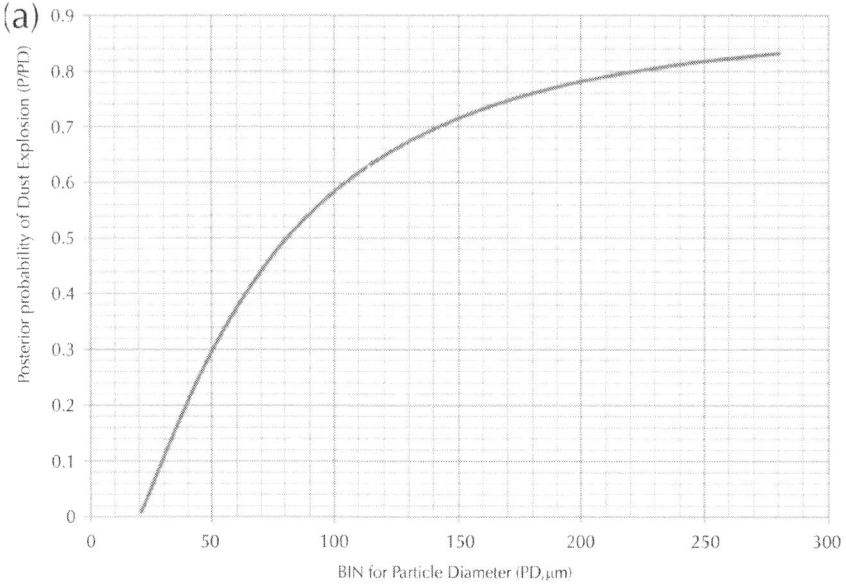

92 Dust Explosions in the Process Industries

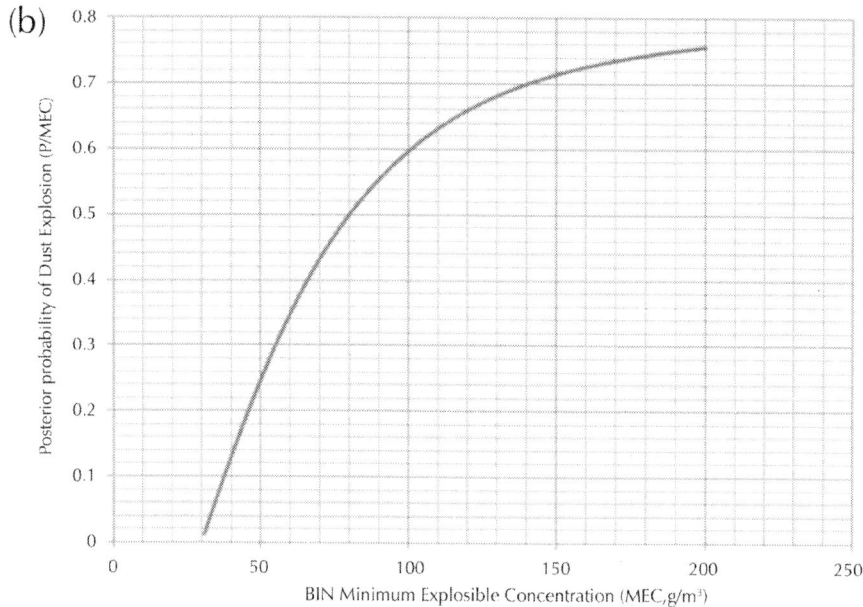

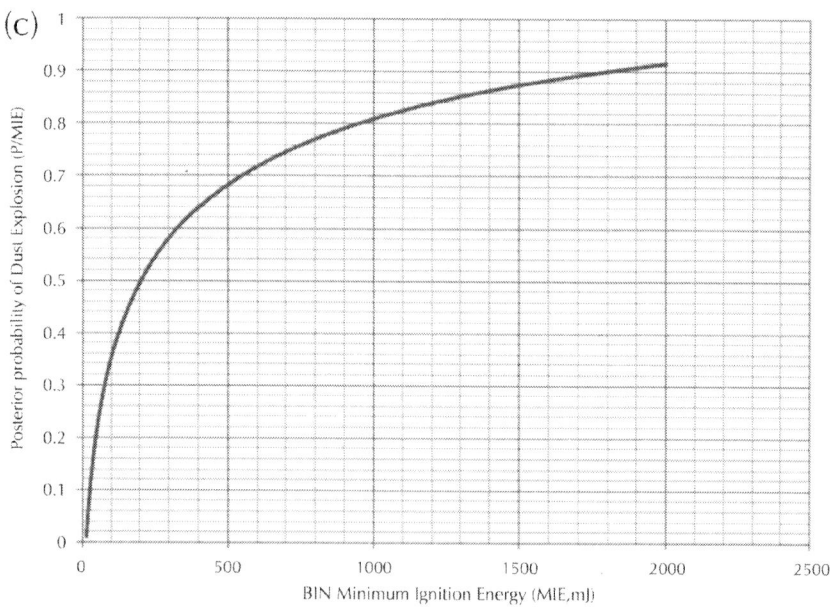

Industry Specific Dust Explosion Likelihood Assessment Model ... 93

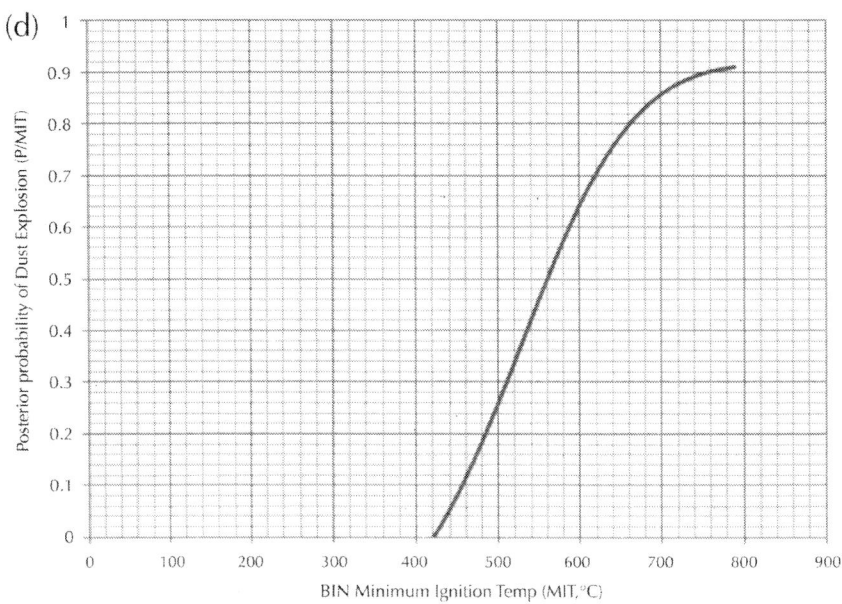

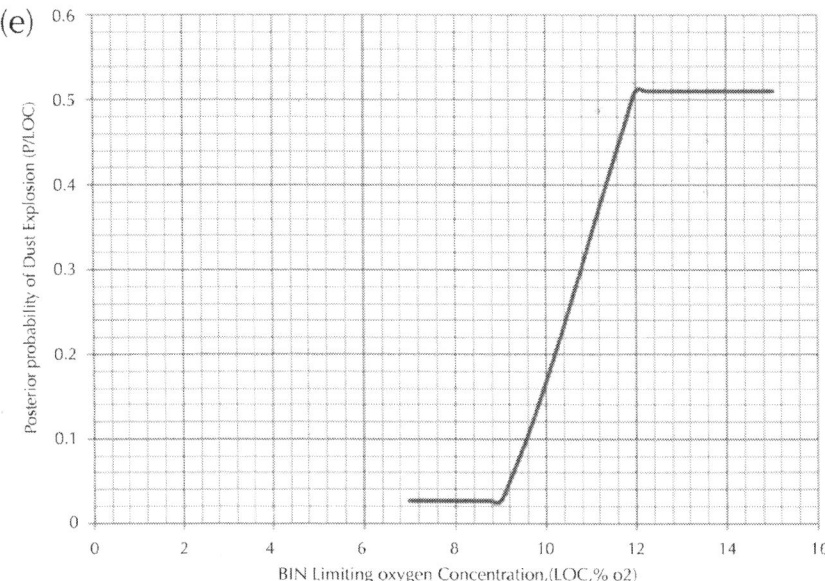

Figure 6: (a) Posterior probability for PD (dust class: plastic, resin and rubber). (b) Posterior probability for MEC (dust class: plastic, resin and

rubber). (c) Posterior probability for MIE (dust class: plastic, resin and rubber). (d) Posterior probability for MIT (dust class: plastic, resin and rubber). (e) Posterior probability for LOC (dust class: plastic, resin and rubber).

Case Study: Dust Explosion in Wool Factory

On 9 January, 2001, at the wool factory Pettinatura Italiana in Vigliano Biellese (BI), at 5:50 p.m. a massive explosion caused the death of three people, the injury of another five and severe facility damage.[5] For over a century, the factory has been devoted to washing, carding, and combing wool. The production cycle which is described includes the extraction from the wool of several types of industrial rejects (e.g. burr) and waste (e.g. noils). A large amount of dust accumulated which was a by-product of the removal of burr from wool during the carding phase. A primary deflagration initiated by some electrical equipment of the lighting system caused a spark or source of heat. The flame front of the primary deflagration propagated quickly and ignited large quantities of dust.

An unusual explosive material was reported as a mixture of vegetable dust, wool fibers, and in-organic substances.[5] According to the estimation the deflagration involved at least 400–500 kg of flammable vegetal and wool fibers, without counting moisture and inert particles.[5] According to the technical reports[2, 5 and 12] and standard material properties, the following plant conditions are determined for the study:

- PD varies from 25 μm to 150 μm,
- MEC range is 125 g/m^3,
- Upper limit for MIT is 450–500 °C,
- Upper limit of MIE is 200 mJ,
- Upper limit for LOC is 10.5 (% of O_2).

Based on the normal operating condition the dust explosion assessment model provides conditional probability and eventually is aggregated to the total probability of explosion. The conditional probabilities, total probability and the generated nomograph are provided in Table 6 and Figure 7 subsequently. During the normal operational condition the total probability of dust explosion in the above industry is 44%. This means that, one out of two operations are susceptible to explosion, which is a high probability for a process facility. The nomograph provides a quick summary of the conditional probabilities and total probability of dust explosion. The occurrence of dust explosion indicates a process upset or standard work procedure breach. Table 6 shows that PD, MEC and MIE make significant contributions to the total probability of dust explosion. The investigation of this accident also notes the ignition source (electrical system of the lighting system) and explosible concentration which caused the initiation of the primary explosion.[5] Based on the quantitative analysis, an industry practitioner, design engineer or administrative personnel can modify the safety measures, work procedures or managerial control to reduce the probability of explosion in normal operating condition. The proposed model provides a simple assessment of dust explosion probability in the normal operating conditions for industries handling dust plastic, resin and rubber.

Table 6: Nomograph (Plastic, Resin and Rubber) Calculation Interface

Influencing parameters	Lower bound	Upper bound	Conditional probability
PD (21–280)	25	150	0.66 (P_{PD})
MEC (31–200)	31	125	0.66 (P_{MEC})
MIT (421–790)	450	500	0.17 (P_{MIT})
MIE (14–2000)	14	200	0.49 (P_{MIE})
LOC (7.25–15)	7.25	10.50	0.25 (P_{LOC})
Total probability of dust explosion			0.44 (P_{Total})

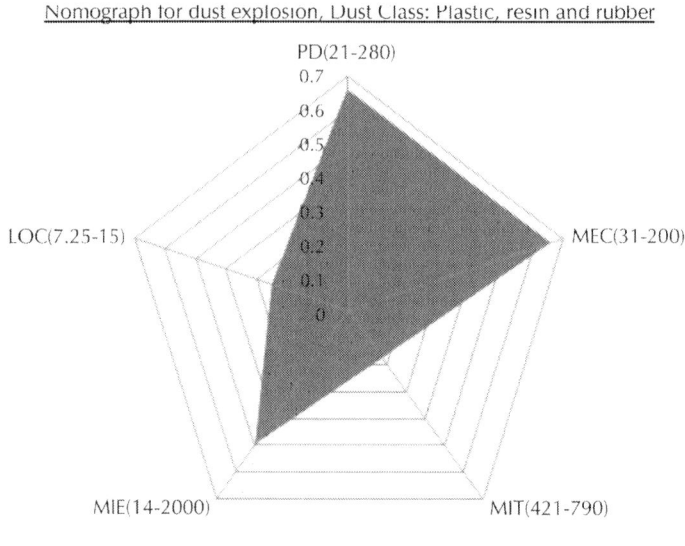

Figure 7: Nomograph (plastic, resin and rubber) for dust explosion.

Dust Class 3: Metal Alloys

In this class the considered dusts are: aluminum powder, bronze powder, manganese, silicon, ferrochromium etc. Industries dealing with such dusts are considered in this class and the result of the statistical analysis for dust explosion parameters of such dust class is listed below. Analyzing the dust explosion parameters and their distribution, the PDFs are determined. These PDFs are integrated over a specific operational range to get the CDFs. The range is determined from the analysis presented in Table 7. The CDFs are plotted in Figure 8. This enables the determination of the conditional probability of dust explosion for a single parameter in metal handling facilities.

Table 7: Metal Alloys: Dust Explosion Parameter Distribution Identification

Dust Explosion Parameter	Best fitted distribution	Estimated distribution parameter	95% data range
PD	Lognormal	$\lambda_{PD} = 3.4, \xi_{PD} = 1.03$	23–250 μm
MEC	Lognormal	$\lambda_{MEC} = 4.8, \xi_{MEC} = 1.034$	31–500 g/m³
MIT	Normal	$\mu_{MIT} = 700.3, \sigma_{MIT} = 156.9$	381–800 °C
MIE	Lognormal	$\lambda_{MIE} = 5.14, \xi_{MIE} = 1.39$	41–250 mJ
LOC	Normal	$\mu_{LOC} = 10.97, \sigma_{LOC} = 2.12$	8.75–12% O_2
P_{max}	Normal	$\mu_{Pmax} = 8.64, \sigma_{Pmax} = 2.29$	5.2–12.4 bar(g)

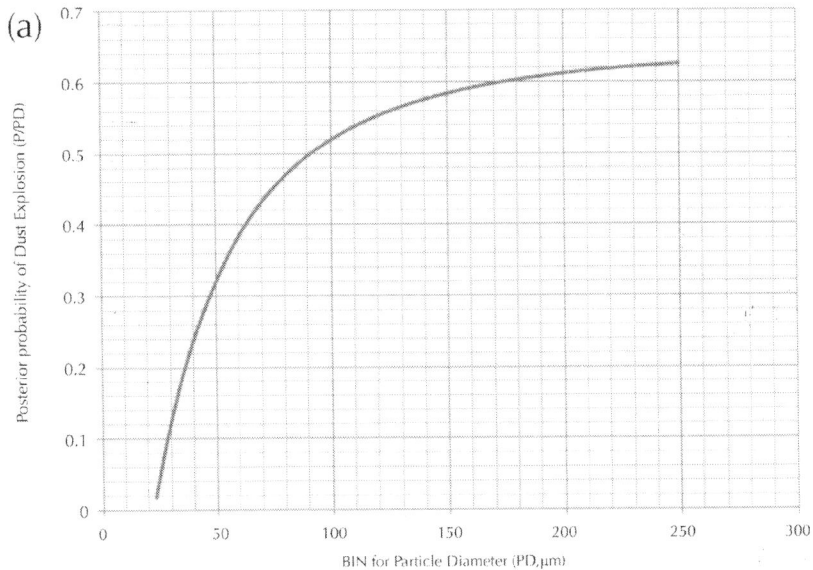

98   Dust Explosions in the Process Industries

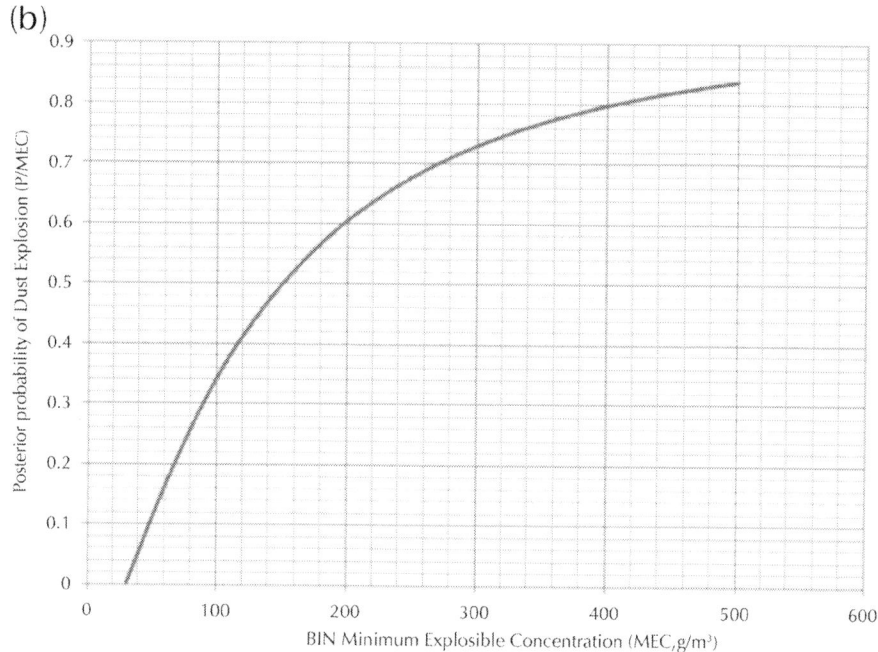

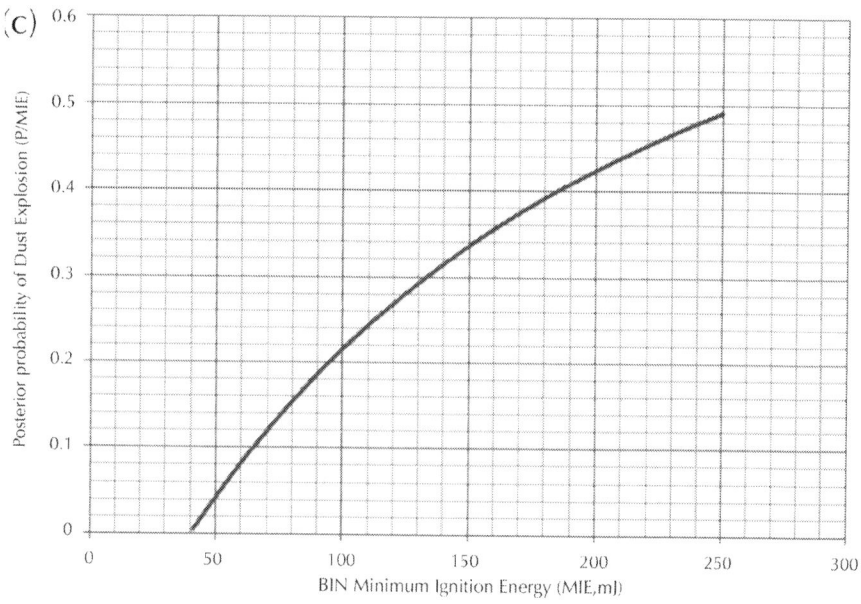

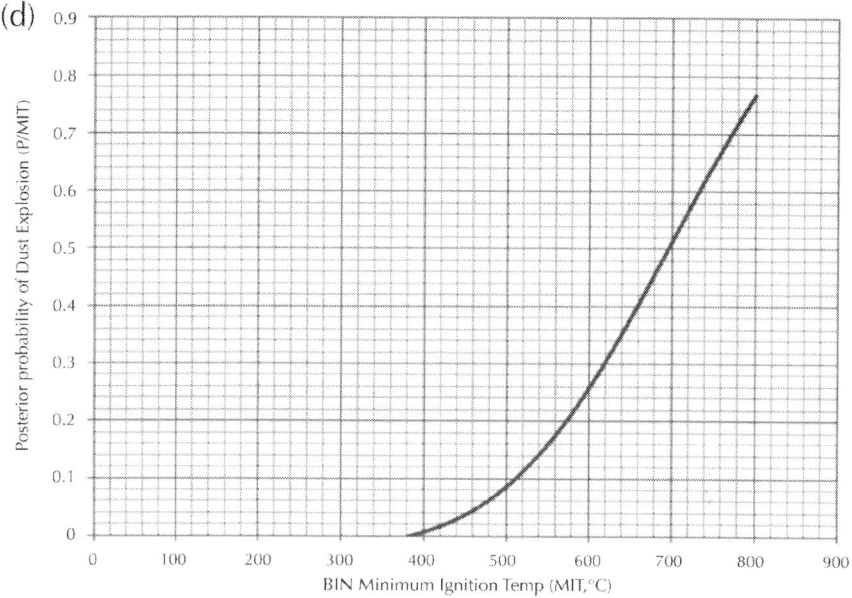

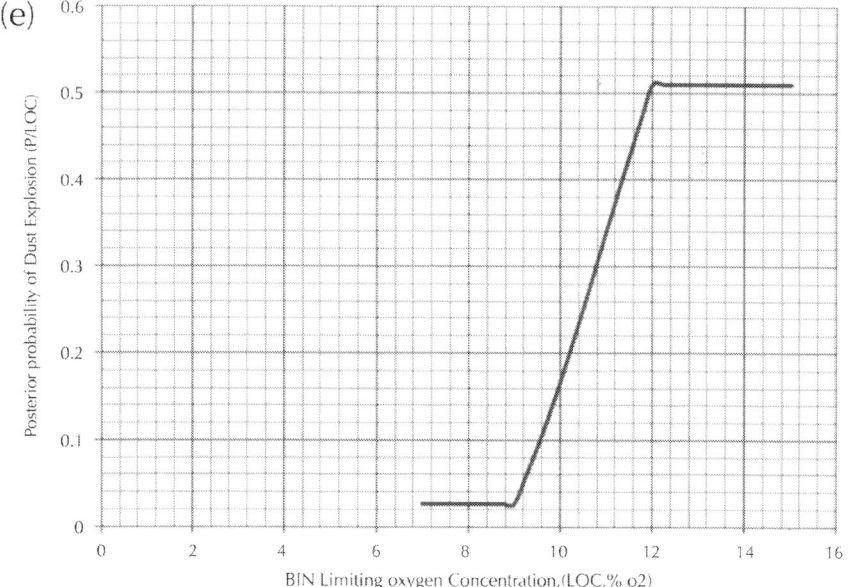

Figure 8: (a) Posterior probability for PD (dust class: metal alloy). (b) Posterior probability for MEC (dust class: metal alloy). (c) Posterior probabil-

ity for MIE (dust class: metal alloy). (d) Posterior probability for MIT (dust class: metal alloy). (e) Posterior probability for LOC (dust class: metal alloy).

Case Study: Aluminum Dust Explosion

At about 8:30 p.m. on Wednesday, October 29, 2003, an aluminum dust explosion and fire occurred at the Hayes Lemmerz International Huntington Inc. (Hayes) facility in Huntington, Indiana.

One employee was engulfed in fire and eventually died and two employees were severely burned. Three employees had minor injuries. The explosion took place in the scrap reprocessing area, near the furnaces in the aluminum casting plant. The explosion completely destroyed dust collection equipment, which was placed outside the building.[4] Equipment inside the building received minor damage. The explosion also lifted a portion of the building roof above one furnace and ignited a fire; insulation and other combustible materials burned for several hours.[4]

Based on the technical reports[2, 4 and 12] and standard material properties the following plant conditions are determined for the study:

- PD varies from 25 μm to 85 μm,
- MEC range is 50 g/m^3,
- Upper limit for MIT is 450–500 °C,
- Upper limit of MIE is 240 mJ,
- Upper limit for LOC is 10.5(% of O$_2$).

For the above operational condition, the model assesses the conditional probabilities, and using the information, the total probability of dust explosion is calculated. The conditional probabilities and total probability of dust explosion are listed in Table 8. The predictive model also provides a nomograph to represent the plant condition in terms of probability of explosion and is depicted in Figure 9. Under normal operational conditions the probability of dust explosion at Hays Lemmerz International was 11%. This means that one out of eight operations is prone to explosion, which is a high probability. Table 8 shows that PD, MEC and MIE make higher contribution to dust explosion occurrence. From the technical report it is also evident that, the dust collector system was not designed properly and the chip system was releasing excess dust, which was unreported before the explosion took place.[4] To reduce the explosion probability in normal operating conditions industries should focus on applicability of fire prevention standards, dust generation and hazard awareness, engineering project management, safety reviews for new and modified systems, and operating and maintenance practices. Adopting such modifications will minimize the probability of explosion in standard operating conditions which will reduce the probability of process upset conditions as well.

Table 8: Nomograph (Metal Alloy) Calculation Interface

Influencing parameters	Lower bound	Upper bound	Conditional probability
PD (23–250)	25	85	0.45 (P_{PD})
MEC (31–500)	31	50	0.11 (P_{MEC})
MIT (381–800)	450	500	0.04 (P_{MIT})
MIE (41–250)	41	240	0.48 (P_{MIE})
LOC (7.25–15)	7.25	10.50	0.23 (P_{LOC})
Total probability of dust explosion			0.11 (P_{Total})

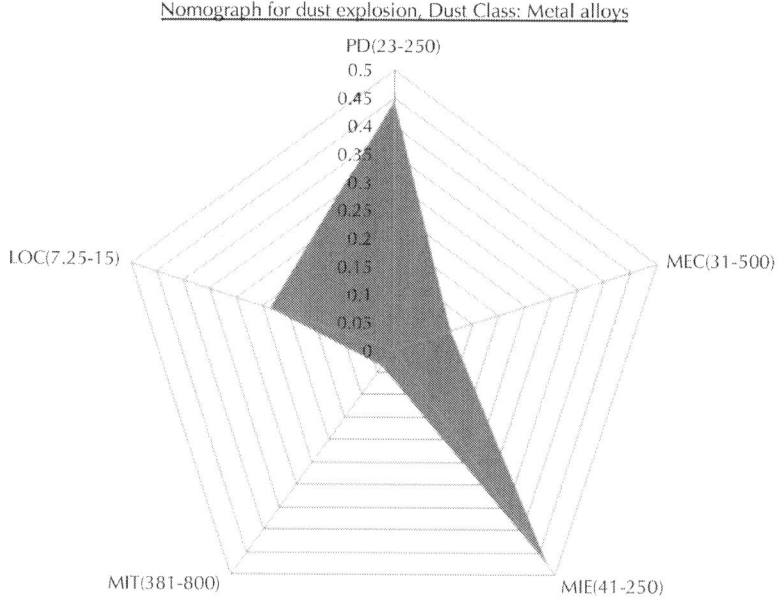

Figure 9: Nomograph (metal alloys) for dust explosion.

DISCUSSION

This paper illustrates a novel approach of estimating dust explosion probability in specific operating conditions. The proposed model assesses dust explosion probability based on the operating conditions. The model is studied with three case studies considering three dust classes: food feed; plastic, resin and rubber; and metal alloys.

The model described in the paper consists of three parts. The first part is monitoring the data. In a specific industry the hazard causing parameters are: PD, MEC, MIE, MIT and LOC. PD, MEC and LOC are monitored and the obtained data are used in the predictive model for dust explosion likelihood assessment. MIT and MIE are material parameters and the specific range depends on the dust material. These values are chosen for the specific dust materials.

After obtaining the necessary data, the second step is the mathematical analysis of data and probability estimation. The model analyses the conditional probability for each parameter and estimates the total probability of dust explosion.

The final step of the model provides a nomograph as a quick estimate of the probability of dust explosion at a given facility. The nomograph is a simplified visual representation of the rigorous analysis process and makes the model easier to interpret. The nomograph provides the assessment of explosion probability in the facility in the base operational condition. Based on the assessment, the designers, engineers and workers can modify their safety measures (e.g. inherent safety, procedural safety, safety management system, etc.) and can formulate effective preventive and mitigatory measures to reduce the probability of explosion to provide a safe working environment.

The proposed model is very convenient and effective in the following ways:

- Easier Implementation: The model consists of systematic structured steps and requires very simple techniques to follow. Moreover, the simplified mathematical steps reduce the data processing and modeling time significantly.
- Easy to Interpret: The outcome of the analysis is presented with a simple visual tool called a "nomograph". Conditional probability and total probability are depicted this way to facilitate easy interpretation of results.
- **Quick Assessing of the Plant's Base Condition**: The model renders a quick assessment of explosion probability in operating conditions for specific dust handling facilities.
- **A Condition Based Approach**: This condition based probabilistic model that will help to assess dust explosion likelihood for a given facility condition.

CONCLUSIONS

The objective of this paper is to discuss a recently proposed model for dust explosion likelihood assessment and to demonstrate its application to real life. Three dust classes are taken into consideration with case studies. Using a conditional probabilistic method enables the model to assess the likelihood of dust explosion for a given operating condition. The model can assess the total probability of explosion in the normal operating condition for a particular process facility. Based on the assessment, the process facility can implement safety measures (e.g. inherent safety, procedural safety, safety management system, etc.) and can develop effective prevention and mitigation strategies to achieve a safer working environment.

Two aspects of the model could be further explored in the future. First, the model can be applied to other dust classes based on chemical compositions that are not covered in this study. A few recommended classifications are:

- Pharmaceuticals, cosmetics and pesticides
- Coal and coal products
- Cotton, wood and peat

Second, copula function requires further testing to develop better dependency scenarios, such as the correlation between P_{max} and PD, P_{max} and LOC.

ACKNOWLEDGEMENTS

The authors gratefully acknowledge the financial support provided by the Natural Sciences and Engineering Research Council (NSERC) of Canada in the form of a strategic grant.

REFERENCES

1. Amyotte, P. R.; Eckhoff, R. K. Dust explosion causation, prevention and mitigation: an overview. J Chem Health Saf, 2010, 17(1), 15–28.
2. NFPA 69. Standard on explosion prevention system. 2008,.
3. CSB. Investigation Report, Dust Explosion, West Pharmaceutical Services, Inc., Report No. 2003-07-I-NC, U.S. Chemical Safety and Hazard Investigation Board. 2004, September,.
4. CSB. Investigation Report. Aluminum Dust Explosion, Hayes Lemmerz International-Huntington, Inc., Report No. 2004-01-I-IN, U.S. Chemical Safety and Hazard Investigation Board. 2005, September,.
5. Piccinini, N. Dust explosion in a wool factory: origin, dynamics and consequences. Fire Saf, 2008, 43(3), 189–204.
6. CSB. Investigation Report: Sugar Dust Explosion and Fire, Report No. 2008- 05-I-GA, U.S. Chemical Safety and Hazard Investigation Board. 2009, September, .
7. Eckhoff, R. Current status and expected future trends in dust explosion research. J Loss Prev Process Ind, 2005, 18(46), 225–237.
8. Kauffman, C. Fuel–air explosion; University of Waterloo Press; Canada, 1982.
9. Abbasi, T.; Abbasi, S. Dust explosions cases, causes, consequences, and control. J Hazard Mater, 2007, 140(12), 7–44.
10. NFPA 68. Guide for venting of deflagration. 2007, .
11. Hassan, J.; Khan, F.; Amyotte, P. R.; Ferdous, R. A model to assess dust explosion occurrence probability. J Hazard Mater, 2013, August, (comments received, 1st revision submitted).
12. Eckhoff, R. Dust explosions in the process industries, 3rd ed. Gulf Professional Publishing; USA, 2003.

Chapter 4

Experiment-based Investigations of Magnesium Dust Explosion Characteristics

Niansheng Kuai, Jianming Li, Zhi Chen, Weixing Huang, Jingjie Yuan, and Wenqing Xu

School of Chemical Engineering, Sichuan University, 29 Wangjiang Road, Chengdu 610065, PR China

ABSTRACT

An experimental investigation was carried out on magnesium dust explosions. Tests of explosion severity, flammability limit and solid inerting were conducted thanks to the Siwek 20 L vessel and influences of dust concentration, particle size, ignition energy, initial pressure and added inertant were taken into account. That magnesium dust is more of an explosion hazard than coal dust is confirmed and quantified by contrastive investigation. The Chinese procedure GB/T 16425 is overly conservative for LEL determination while EN 14034-3 yields realistic LEL data. It is also suggested that 2000–5000 J is the most appropriate ignition energy to use in the LEL determination of magnesium dusts, using the 20 L vessel. It is essential to point out that the overdriving phenomenon usually occurs for carbonaceous and less volatile metal materials is not notable for magnesium dusts. Trends of faster burning velocity and more efficient and adiabatic flame propagation are associated with fuel-rich dust clouds, smaller particles and hyperbaric conditions. Moreover, Inerting effectiveness of $CaCO_3$ appears to be higher than KCl values on thermodynamics, whereas KCl represents higher effectiveness upon kinetics. Finer inertant shows better inerting effectiveness.

INTRODUCTION

Magnesium is an excellent engineering material due to its light weight and high mechanical strength. As one of the most important magnesium products, magnesium powders are widely used in metallurgy, aeronautics, fireworks, painting and chemical industries. Nevertheless, magnesium is combustible and explosive. There could be severe hazards in production, storage, transportation and handling. While the concentration of suspended magnesium dust in a confined space exceeds the lower explosion limit (LEL), explosions will occur with the effect of ignition sources. In fact, dust explosions represent significant damages in industries. As far as China is concerned, accidents happened with losses of human lives and destruction of industrial facilities. For example, a severe magnesium dust explosion described by Zhang, Jiang, and Zheng (2005) resulted in the death of five people in 1978. Another case described by Kuai, Li, and Chen (2010) caused one death in 2008.

Safety handling of magnesium powders first requires data on explosion hazards. Thus it is urgent to increase the understanding of magnesium dust explosion characteristics and explosion mitigation.

Several studies were carried out on this field and data in previous articles mostly focused on the determination of explosion sensitivity parameters and explosion severity characteristics of magnesium dust (Li et al., 2009a,Li et al., 2008 and Nifuku et al., 2007). Unfortunately, (i) few studies included a systematic study showing influences of dust concentration and particle size, above all, influences of ignition energy or initial absolute pressure are hardly mentioned. (ii) The comparison between Chinese and European LEL determination procedures has not been practiced and the most appropriate ignition energy for LEL determination of magnesium dust has not been discussed yet. (iii) Inerting of magnesium dust with chloride (widely used as the powder extinguishing agent for light metal fire) has not been practiced yet.

This work aims to present data about the overall characteristics of magnesium dust ignitability and explosibility. (i) Information about the flammability limit (i.e. LEL) and severity characteristics (maximum explosion pressure p_{max}, maximum rate of pressure rise $(dp/dt)_{max}$ and combustion time t_c) was given. (ii) Influences of some factors, like dust concentration, particle size, ignition energy, initial absolute pressure, and added inertant, were analyzed. (iii) The most appropriate ignition energy for LEL determination of magnesium dust was presented. (iv) The inerting effectiveness of different inertants was compared.

MATERIALS AND EXPERIMENTAL SETUPS

Materials and Apparatus

The tested magnesium powders with 99% purity provided by CNPC POWDER were produced by atomization. Their particle size distributions, illustrated in Table 1, were determined using digitized video images by a microscope. Powders of $CaCO_3$ and KCl were chosen as inertants. Two samples of $CaCO_3$ were prepared by sieving and their particle size distributions were nominal minus 400 and 3000 mesh respectively. The particle size of KCl powder was nominal minus 120 mesh. All the samples were systematically dried before handing. Experiments were performed in the well-known 20 L sphere developed by Siwek (1996) initially. The vessel is an explosion resistance hollow sphere made of stainless steel with a volume of 20 L in accordance with the recommendations of European standard EN 14034 (CEN/TC305, 2004a, CEN/TC305, 2004b and CEN/TC305, 2004c), ASTM standard E 1226 (ASTM, 2007) and Chinese standard GB/T 16425 (MCI, 1996). Dust clouds in the vessel are ignited by pyrotechnical ignitors. The ignition energy depends on the mass of the pyrotechnical composition which consists of zirconium, barium nitrate and barium dioxide by the ratio of 4:3:3 by weight.

Table 1: Granulometric properties of magnesium powder samples

	d_{10} ^{a} (µm)	d_{50} (µm)	d_{97} (µm)
Sample A	24.9	54.5	129.9
Sample B	12.3	22.4	88.8
Sample C	4.1	7.5	23.9

A d_x is defined as the size at which $x\%$ of the particles has smaller diameters.

Flammability limit

The LEL is the minimum dust concentration at which an explosion occurs and it is the most important flammability limit parameter. Keeping the dust concentration below LEL is considered as an inherent safety approach (Amyotte et al., 2009 and Lutz, 1997). In the present work, flammability limit studies were carried for each sample according to the specifications of EN 14034-3 and GB/T 16425 respectively. EN 14034-3 demands ignitors with 2000 J and presents an explosion criterion: $p_{max} \geq 0.5$ bar. Nevertheless,

GB/T 16425 prefers 10000 J ignition source and requires a higher explosion criterion: $p_{max} \geq 1.5$ bar. The highest dust concentration at which the p_{max}-value is just below the criterion in three consecutive tests was regarded as the LEL.

Explosion Severity

The explosion severity parameters measured in the Siwek 20 L are p_{max}, $(dp/dt)_{max}$ and t_c. Values of p_{max} are typically related to the thermodynamics concerned with the amount of heat liberated during combustion, whereas $(dp/dt)_{max}$ and t_c are concerned with the rate at which the reaction heat is liberated. Influences of dust concentration, particle size, ignition energy and initial pressure were taken into account. Experiments were performed for each magnesium sample, and the dust concentration ranged from 40 to 1500 g m^{-3}. Ignitors with energies of 1000, 2000, 2500, 5000, 10000 and 20000 J were applied, and initial absolute pressure ranged from 1.0 to 1.6 bar. All the tests were performed in three replications at least. For readability purpose, the numerical mean deviations have been indicated by means of error bars or scatter diagrams in each figure.

Inerting of Magnesium Dust

Inerting, i.e. to mix the combustible dust with inert substances, is also considered as an inherent safety approach (Amyotte, 2006). This principle of inerting has been practiced in underground coal mining where rock dusts are mixed with coal dust to prevent explosion. Other applications do exist (Chatrathi and Going, 2000, Dastidar and Amyotte, 2002a and Mintz et al., 1996). In the present work, $CaCO_3$ and KCl powders were chosen as inertants. The wide use of rock dusts consist of limestone in coal mining industries gives the idea that $CaCO_3$ can be applied to prevent or mitigate magnesium dust explosions. KCl shows a potential of inerting because it is widely used as the powder extinguishing agent for light metal fire. The combustible-inert mixtures were well-mixed before testing. The mixing ratio of inertant ranged from 0 to 80%. All of the inerting studies were performed using 10000 J ignitors and at 1.0 bar initial absolute pressure.

Reaction Mechanisms

Two alternative mechanisms have been proposed for magnesium dust explosions: the vapor-phase and the heterogeneous mechanisms. The present work lends support for both mechanisms. Dreizin and Hoffmann, 1999 and Dreizin and Hoffmann, 2000 proposed a systematic mechanism to describe the magnesium dust flame propagation. The observed flame structure contains preheat and combustion zones. They suggested that Mg–O produced in the vapor-phase flame is not the primary source of the Mg–O coating found on the burnt particle surfaces. The heterogeneous mechanism of forming irregular oxide coatings is via the formation of an Mg–O solution followed by an exothermic phase separation occurring within the burning particles, consistent with the single particle combustion studies (Dreizin, 2000, Dreizin, 2003 and Dreizin et al., 2000). The oxygen dissolution in the molten particles begins simultaneously with the vapor-phase combustion but later on, after the entire vapor has been oxidized, becomes the primary combustion mechanism.

RESULTS AND DISCUSSION

Influence of Dust Concentration

Experiments were performed for the dust concentration ranging from 40 to 1500 g m^{-3}, using ignitors of 10000 J. Take the case of sample B, evolution of p_{max} was plotted as a function of dust concentration in Fig. 1. The typical pattern of variation with dust concentration was observed and especially the increase of p_{max} for lean dust concentration up to 1000 g m^{-3}, followed by a decrease of p_{max} for highly loaded dust clouds. Similar evolutions were investigated for samples A and C.

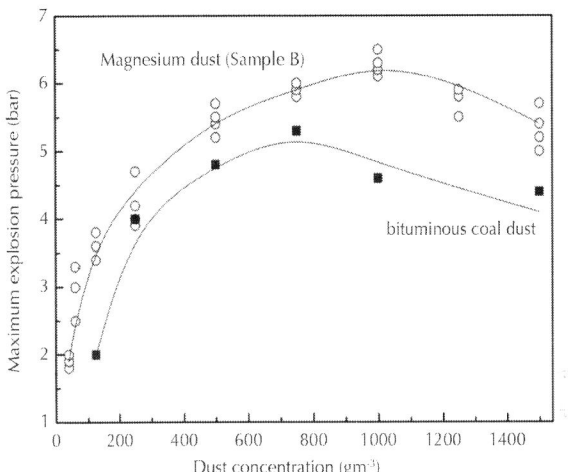

Figure 1: Evolutions of p_{max} with dust concentration, for magnesium particles with a median diameter of 22.4 µm and bituminous coal particles with a median diameter of 27.6 µm.

The dust concentration 1000 g m^{-3} where the evolution of p_{max} peaks is defined as the worst-case concentration. Nevertheless, the stoichiometric concentration for Mg/O$_2$ combustions, relatively to the oxygen content in the atmosphere, could be estimated by the following equation:

$$Mg(s)+O_2 \cdot {}_5O(g)=MgO(s) \tag{1}$$

For initial absolute pressure of 1.0 bar, it equals to about 450 g m^{-3}, which is about two times lower than the worst-case. The outstanding deviation between the stoichiometric and worst-case was also observed for aluminum dusts. Although the deviation is well known, it is essential to point out that deviations for magnesium and aluminum are obviously higher than those for some other metal dust like titanium, iron and niobium (Cashdollar & Zlochower, 2007). This behavior gives the idea that the Mg (Al)/N$_2$ reaction somewhat need consideration. If the Mg/N$_2$ reaction is taken into account, the combustible concentration region is actually expanded. Nevertheless, the confirmed inerting effectiveness of nitrogen on magnesium dust explosion (Li et al., 2009b, Li et al., 2009a and Nifuku et al., 2007) means that the domination of Mg/O_2 reaction is pointed out, and thereby synthesizing magnesium nitride by combustion may occur in the oxygen deficient atmosphere.

Note that the decrease of p_{max} was observed for highly loaded dust clouds. Explosion pressure drops to 5.3 bar at dust concentration of 1500 g m^{-3}, which corresponds to a yield of 84% of the worst-case. It is obvious that some of particles cannot be completely oxidized. A sequence of micrographs reveals that explosion products vary significantly from lean to rich mixtures (Fig. 2). The irregular shape of burned particles was observed at concentration of 125 g m^{-3}, whereas a great number of unburned particles were observed at 1500 g m^{-3}. The absence of surface oxide layer on the partially burned particles means that the phase separation (Dreizin et al., 2000) does not occur for these particles because they have been quenched before the solution became saturated (Dreizin & Hoffmann, 2000). This behavior is related to the oxygen deficiency within highly loaded dust clouds. Furthermore, less efficient heat transfers within highly loaded dust clouds may be an alternative interpretation.

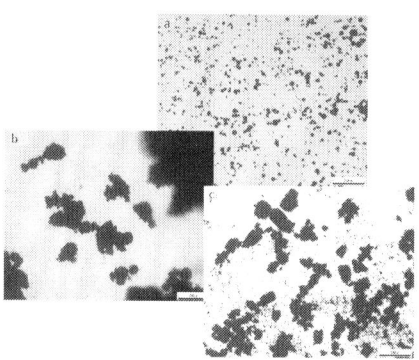

Figure 2: Microscope observations for sample C: a) original particles; b) explosion products at the concentration of 125 g m^{-3}; c) explosion products at the concentration of 1500 g m^{-3}.

Moreover, the severe overpressure of magnesium dust explosion was emphasized in comparison with coal dusts. Values of p_{max} for bituminous coal dust (d_{50} = 27.6 μm) obtained by authors' group (Gao et al., 2010 and Li et al., 2009b) versus those for magnesium were plotted in Fig. 1. That magnesium dust is more of an explosion hazard than bituminous coal is confirmed and quantified.

Evolution of $(dp/dt)_{max}$ was plotted as a function of dust concentration in Fig. 3a. The p_{max} peaks at the concentration of 1000 g m^{-3}, whereas the evolution of $(dp/dt)_{max}$ reaches a fairly constant peak level in the range from 1250 g m^{-3} and upwards. The stabilization of $(dp/dt)_{max}$ at concentrations higher than the worst-case was also identified for aluminum dust (Dufaud, Traore, Perrin, Chazelet, & Thomas, 2010). Nevertheless, no reappearance about the mentioned stabilization was identified for cork dust (Pilao, Ramalho, & Pinho, 2006), silicon dust (Matsuda, Yashima, Nifuku, & Enomoto, 2001) and graphite dust (Denkevits & Dorofeev, 2006). Consequently, the stabilization of $(dp/dt)_{max}$ as well as burning velocity within highly loaded dust clouds means that the evidence for Mg/N$_2$ reaction in the oxygen deficient atmosphere is somewhat found again.

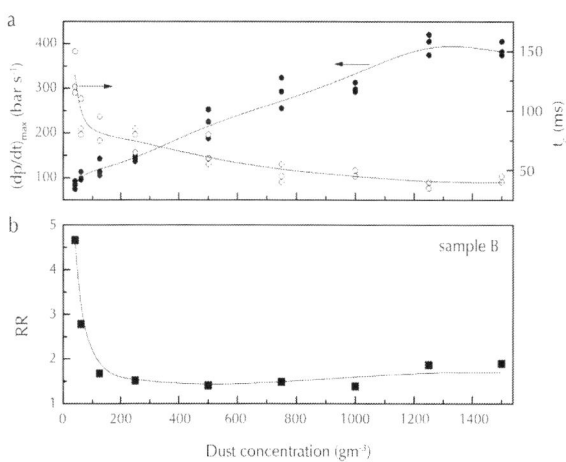

Figure 3: Evolutions of $(dp/dt)_{max}$, t_c and RR with dust concentration, for magnesium particles with a median diameter of 22.4 μm.

In the same way, the other kinetic parameter t_c, which is the duration between ignition and maximum overpressure achievement, drops from 130 to 40 ms by varying dust concentration from 40 to 1500 g m^{-3} (Fig. 3a). It is consistent with the previous assessments of $(dp/dt)_{max}$.

The ratio between p_{max} and t_c gives the average rate of pressure rise $(dp/dt)_{aver}$. Typical pressure–time curves were plotted in Fig. 4. As can be seen, $(dp/dt)_{aver}$ presents an ideal pressure development during the combustion process. Actually, at the early stage of dust explosions, the burning velocity is slower than the average one because particles need preheating prior to combustion, which can be the reason why the flame propagation undergoes an ineffective period, that is, the induction time (t_i). At the end of pressure development, the top pattern of curve means that the slowdown of burning velocity is verified. This behavior is related to the heat losses toward the chamber wall. Consequently $(dp/dt)_{max}$, the peak value obtained from the pressure–time curve, is always higher than the corresponding $(dp/dt)_{aver}$.

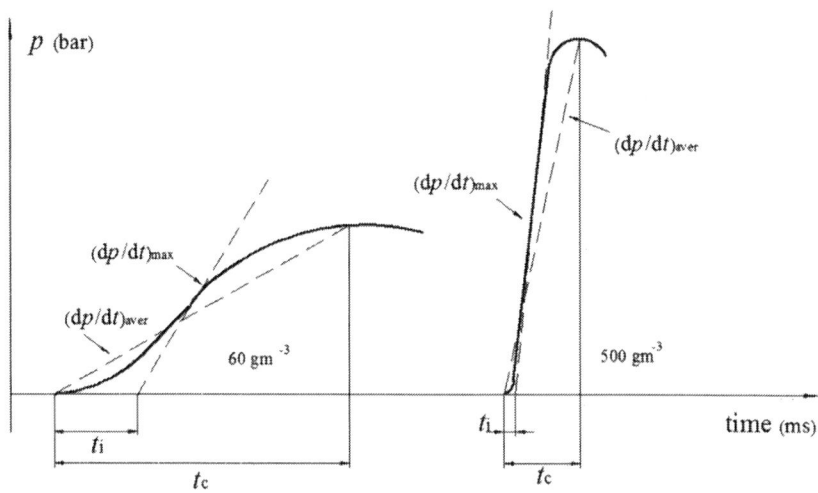

Figure 4: Pressure–time curves at dust concentrations of 60 and 500 g m^{-3}, for magnesium particles with a median diameter of 22.4 μm.

To seek for the relations between $(dp/dt)_{aver}$ and $(dp/dt)_{max}$, a dimensionless parameter rate ratio was introduced, defined as RR=$_{[(dp/dt)max-(dp/dt)aver]/(dp/dt)aver}$. It obvious that RR decreases significantly as dust concentration is raised, followed by a stabilization (Fig. 3b). The decrease of RR gives the idea that the deviation between $(dp/dt)_{max}$ and $(dp/dt)_{aver}$ is reduced. This behavior can be related to the more efficient and adiabatic combustion process. The corresponding evidence was plotted in Fig. 4. As can be seen, both the induction time and heat losses are reduced as dust concentration varies from 60 to 500 g m^{-3}. The fuel–lean mixture undergoes worse inter-particle heat transfer and more heat losses while the stabilization of combustion within fuel-rich mixtures is verified (Goroshin, Fomenko, & Lee, 1996). Moreover, the stabilization of RR is consistent with the proportion between $(dp/dt)_{aver}$ and $(dp/dt)_{max}$ (Proust, Accorsi, & Dupont, 2007).

Influence of Particle Size

Explosion Severity

Evolutions of $(dp/dt)_{max}$ were plotted as a function of dust concentration for various samples using ignitors of 10000 J (Fig. 5). Results show that for fuel–lean mixtures, $(dp/dt)_{max}$ is somewhat independent of particle size. Above concentration of 125 g m^{-3}, a rapid decrease of $(dp/dt)_{max}$ with the rise of particle size is observed. The decrease of $(dp/dt)_{max}$ could be interpreted by a strong decrease of specific surface area, which impacts the reaction kinetics. According to the above-mentioned vapor-phase and heterogeneous mechanisms, efficiencies of evaporation and oxygen absorption are dependent of particle dimensions. The rate of magnesium particle heating is inversely proportional to the square of particle diameter (Dreizin & Shoshin, 2003). Hence the ignition delay, the duration required to heat a particle from the initial temperature to the ignition temperature, is shortened as particle size is diminished. While flame propagates through magnesium dust clouds, the noticeable trend of a wider combustion zone associated with larger particles (Dreizin & Hoffmann, 1999) is due to slower flame propagation for larger particles (Dreizin and Hoffmann, 2000 and Dreizin and Shoshin, 2003). The time of combustion zone to pass through a given volume is found to increase with particle size (Dreizin & Hoffmann, 1999). In addition, that the flame velocity in the magnesium–air aerosol combustion varies with particle dimension (Dreizin & Shoshin, 2003) once again provides an evidence for trends of faster burning velocity and more efficient flame propagation associated with smaller particles. Higher reactivity for finer metal particles is also ascribed to the reduced minimum ignition temperature, minimum ignition energy and activation energy (Li et al., 2008 and Rai et al., 2006). Moreover, the significance of particle size was also underlined for other metal materials (Dufaud et al., 2010, Huang et al., 2009 and Tang et al., 2009).

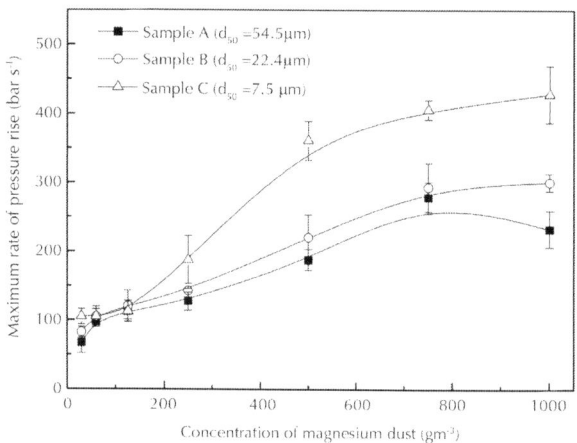

Figure 5: Evolutions of $(dp/dt)_{max}$ with dust concentration for various particle sizes.

It should be pointed out that magnesium oxidation occurs in the atmosphere prior to ignition. Considering the thickness of oxide layer remains constant, the active content is raised with increasing particle size (Li et al., 2008). The lack of reactivity for larger sample is somewhat partially balanced by its more active content under the fuel–lean condition, which is the reason why the differences between $(dp/dt)_{max}$-values of finer and larger particles are inconspicuous for fuel–lean mixtures.

Evolutions of t_c were plotted as a function of specific surface area characterized by the inverse of median diameter for various dust concentration conditions (Fig. 6). This indicates that t_c increases with particle size, consistent with the previous assessments of a link between combustion kinetics and $(dp/dt)_{max}$. Moreover, the decrease of overall reaction time induces fewer heat losses toward the vessel wall. Note that the evolution of t_c is not linear. Maybe this is due to the broad distributions of particle size. The presence of a fine particle size fraction always enhances the flame propagation (Cashdollar, 2000). Agglomeration, reduces the effective specific surface area, is a general trend for fine particles in particular when particle diameter below 10 μm (Eckhoff, 2003) and thereby it might be an alternative reason (Abbasi & Abbasi, 2007). The evidence for agglomeration was provided by the microscope measurement (Fig. 5a). Furthermore, this behavior is because the overall combustion process transmits to the reaction-controlled mode because the rate of evaporation is sufficiently fast for fine particles (Pilao et al., 2006).

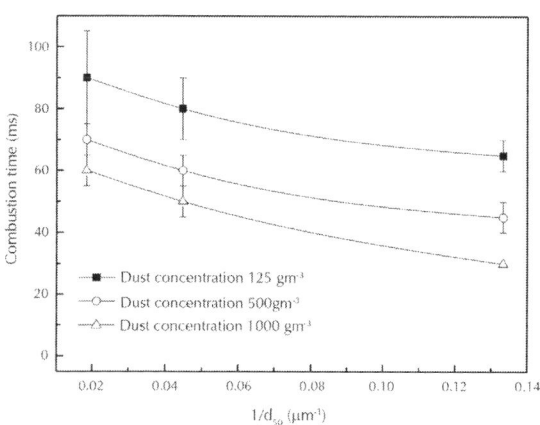

Figure 6: Evolutions of t_c with the inverse of median diameter for various dust concentrations.

In addition to these results, data of p_{max} measured for various samples were plotted in Fig. 7. Results show that for poorly loaded dust clouds, p_{max} is somewhat independent of particle size. Above concentration of 125 g m^{-3}, a rapid decrease of p_{max} with the rise of particle size is observed. Considering that p_{max} is typically related to the thermodynamics, the decrease of p_{max}-value means that the decreasing amount of heat liberated by the overall combustion process is verified. Interpretations are given as follows. On one hand, ignition strengths required to ignite the magnesium particles increase with particle size (Li et al., 2008) and larger metal particles present higher activation energy (Rai et al., 2006). Considering that p_{max} includes the effect of ignitors (Detailed information will be discussed in the section of 3.3 Influence of ignition energy'), it follows that a great amount of ignition energy is dedicated to the ignition process and a limited one to flame propagation for larger particles. Consequently, more energy released by finer particles benefits from less consumption during ignition phase. On the other hand, fewer heat losses associated with shorter combustion duration give the idea that, as the particle size is diminished, the more efficient and adiabatic is the combustion process.

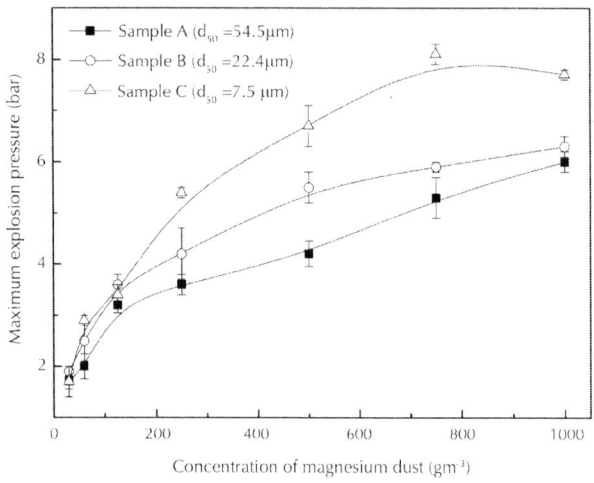

Figure 7: Evolutions of p_{max} with dust concentration for various particle sizes.

The differences between p_{max}-values of finer and larger particles are inconspicuous for poorly loaded dust clouds because the lack of reactivity for larger particles is somewhat partially balanced by its more active content under the fuel–lean condition, consistent with the previous assessment of $(dp/dt)_{max}$.

Flammability limit

Applying the specifications of EN 14034-3 to the obtained results with the three tested particle sizes; the LEL is obtained for each one (Fig. 8). This figure shows that a pronounced increase of LEL can be observed with the rise of particle size, which indicates that finer particles represent greater explosibility. This is because larger particles will undergo less rapid heating to the extent that sample does not explode until a high enough dust concentration has reached.

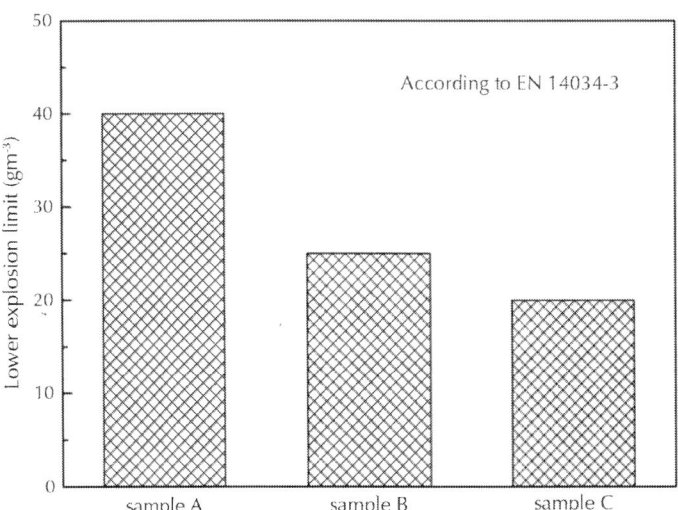

Figure 8: LEL for various particle sizes, according to EN 14034-3.

Moreover, LEL-values in present work are lower than those reported by Nifuku et al. (2007). The deviation is probably due to different ignition energies. In fact, the strength of ignition source has a great influence on the measured LEL (Cashdollar and Chatrathi, 1993, Chawla et al., 1996 and Going et al., 2000) (Detailed information about this aspect will be discussed in the section of '3.3 Influence of ignition energy'). Furthermore, some reasonable theories were established to predict LEL (Eckhoff, 2003). According to Jaeckel model which is successful in the prediction of aluminum's LEL (Mittal, 1997), the predicted LEL of magnesium dust is about 25 g m^{-3}, somewhat in agreement with authors' experimental data. Unfortunately, the model does not take the particle size into account.

Influence of Ignition Energy

Ignition Behavior of Ignitors

In the ignition process, pyrotechnic ignitors explode rapidly with a very strong flame spread. The convection of flame products with high temperature acting like a jet not only heats the magnesium particles but induces some degree of turbulence in the testing vessel (Zhen & Leuckel, 1997). This kind of volumetric ignition is obviously much stronger than an electrical spark. Heat liberated by pyrotechnical ignitors would directly lead to a pressure rise $p_{ignitor}$. Nevertheless, the presence of $p_{ignitor}$ is always under the mask of p_{max}. In the present work, $p_{ignitor}$ was picked out by determining pressure rise due to ignitors themselves without adding magnesium samples. Fig. 9a shows that $p_{ignitor}$ is proportional to ignition energy $E_{ignitor}$ and the fitting equation is given as follows:

$$p_{ignitor} = 1.08 \times 10^{-4} \times E_{ignitor}$$

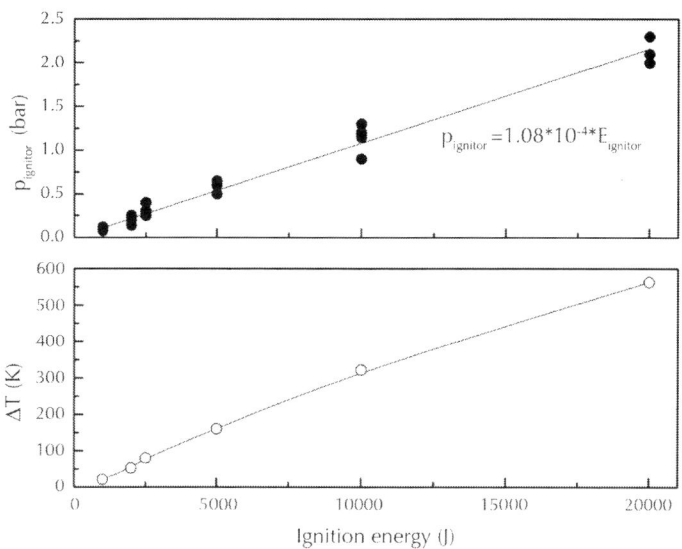

Figure 9: Evolutions of $p_{ignitior}$ and ΔT with ignition energy.

Values of $p_{ignitor}$ are similar to those measured by Cashdollar and Chatrathi, 1993 and Mintz et al., 1996 and Pilao et al. (2006).

The effective energy transferred from ignitors to the atmosphere E_{eff} and the temperature rise due to ignitors ΔT can be estimated thanks to the first principle of the thermodynamics (Proust et al., 2007). It can be calculated that 40–60% of the ignition energy is transferred to the atmosphere and the remaining is lost in solid residues. The ignition strength significantly affects the initial temperature in the 20 L vessel (Fig. 9b) while the initial temperature appears to be stabilization in the 1000 L vessel (Proust et al., 2007). The rise of initial temperature would significantly shorten the ignition delay and increase the evaporation rate (Dreizin & Hoffmann, 1999). In addition, the augmenting ignition strength results in the rise of ignition zone (Cashdollar and Chatrathi, 1993 and Zhen and Leuckel, 1997). Consequently, pyrotechnic ignitors will accelerate the burning velocity as a result from both the temperature rise and the volumetric ignition phase. The effect will be quantitatively evaluated by the following researches.

Explosion Severity

Different actions of magnesium dust explosions under various ignition energies were observed. Date obtained with dust concentration of 500 g m^{-3} was picked out for a typical discussion (Fig. 10). Explosions initiated with 1000 J present a slow burning velocity. Due to the weak activation a small part of dust cloud is ignited, which to a larger extent would allow a self-sustained dust flame to propagate away from the ignition zone toward the vessel wall in terms of the simple spherical flame propagation (Evans, 1994). As the ignition energy is raised, the volume of ignition zone augments and thereby the effective flame front stretches. Moreover, the increased temperature rise can promote the melting and evaporation. Hence, the burning rate is certainly accelerated by 2500 and 5000 J ignitors. Alternatively, the turbulence induced by strong ignitors also promotes the burning velocity because it gives rise to mixing of the hot burned and burning parts of the cloud with the unburned parts while dust cloud is burning (Eckhoff, 2003). In the case of 5000–10000 J ignitors, the ignition zone is so large that a limited size is left to the self-sustained flame propagation. The stabilization of kinetic parameters with ignition strength varying from 5000 to 10000 J mean that the effect of ignition source is slightly reduced because the flame fronts contact with the chamber wall far before the real end of the explosions (Zhen & Leuckel, 1997). In the case of 20000 J ignitors, the heterogeneous mechanism may help to interpret the result. The corresponding initial temperature is close to the molting point of magnesium 650 °C. Such level of preheating may lead to the oxygen solution on the particle surface far before the flame front arrives, which will significantly accelerate the surface heterogeneous oxidation of the Mg–O solution. Apparently the ignition is almost volumetric with much of particles starting to burn simultaneously.

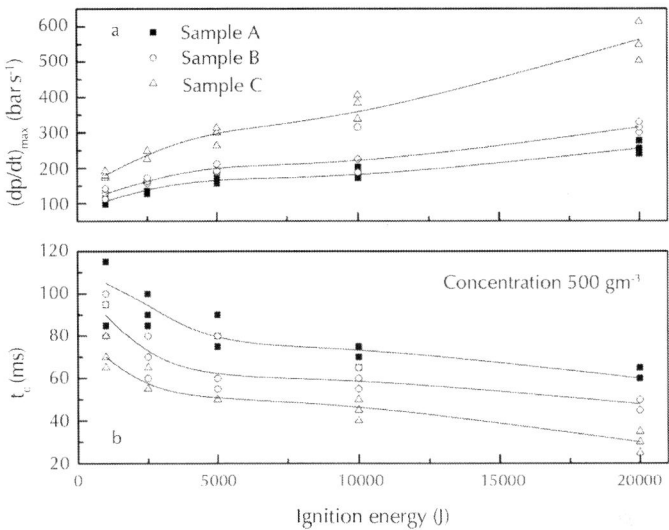

Figure 10: Evolutions of $(dp/dt)_{max}$ and t_c with ignition energy for various particle sizes.

It is well know that p_{max} increase with ignition energy (Pilao et al., 2006 and Zhen and Leuckel, 1997). This is because p_{max} is measured by the total energy release including $p_{ignitor}$. Take the case of dust concentration 400 g m⁻³, next to the stoichiometric concentration, p_{max} increases linearly with ignition energy (Fig. 11a), according to the following equation:

pmax=1.08×10⁻⁴×Eignitor+4.05

The relation between equation (3) and equation (2) gives the idea that p_{max} is a combination of pressure rise due to ignitors and magnesium dust combustion. A dimensionless parameter pressure ratio is introduced to subtract the impact of ignitors. Pressure ratio is defined as PR=(pmax+pi−pignition)/pi, where p_i is the initial absolute pressure (1.0 bar). The good stabilization of PR reveals that the amount of heat liberated during explosion process is relatively independent of ignition energy (Fig. 11b). Thus the overdriving phenomenon which usually takes place for carbonaceous materials and some less volatile metal is not notable for magnesium dusts. This is ascribed to both the high volatility of magnesium (Cashdollar,

1994 and Hertzberg et al., 1992) associated with the vapor-phase combustion and the low melting point associated with the surface heterogeneous oxidation.

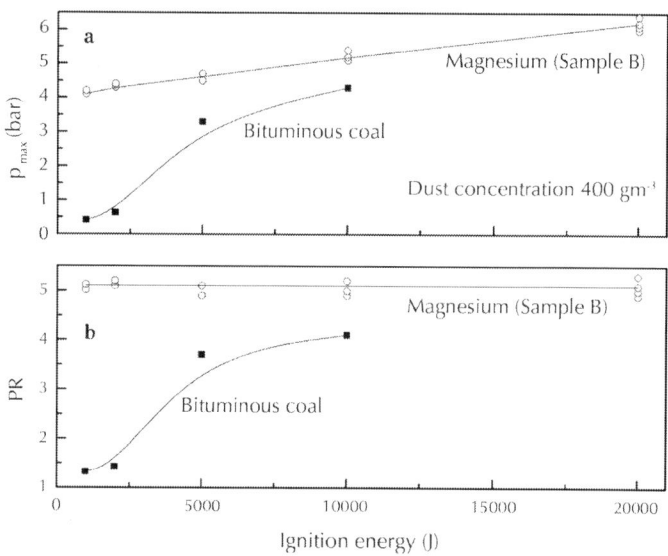

Figure 11: Evolutions of p_{max} and PR with ignition energy, for magnesium particles with a median diameter of 22.4 μm and bituminous coal particles with a median diameter of 29.4 μm.

To round off the discussion, a contrastive investigation was carried out on bituminous coal dust (d_{50} = 29.4 μm). The increase of PR with ignition strength means that the overdriving of coal–air system is verified (Fig. 11b). According to the flame propagation mechanism of coal dusts (Hertzberg et al., 1982 and Hertzberg and Zlochower, 1990), the flame propagation starts with the devolatilization followed by a gas-phase combustion. Due to the low volatility, the volatile yields of coal dusts are related to the strength of ignition sources (Chawla et al., 1996). The significant overdriving for coal dusts was also underlined by Cashdollar and Chatrathi (1993) and Going et al. (2000). Detailed information about this aspect can be found in authors' previous studies (Gao et al., 2010 and Li et al., 2009b).

Flammability limit

Values of LEL measured according to GB/T 16425 versus data according to EN 14034-3 were illustrated in Fig. 12. It can be found the GB/T 16425 method is somewhat conservative. It is essential to stress that their different criteria can be replaced by a new modified criterion: PR ≥ 1.3. As can be seen, the LEL depends on the strength of ignition energy. Nevertheless, the realistic LEL should be determined in conditions where it is independent of the ignition strength (Cashdollar and Chatrathi, 1993, Going et al., 2000 and Hertzberg et al., 1981). Flammability limits vary significantly with ignition energy in the 20 L vessel (Chawla et al., 1996, Dastidar and Amyotte, 2002a and Myers, 2008), whereas the 1000 L vessel usually shows independence (Cashdollar and Chatrathi, 1993 and Going et al., 2000). This is partially because the volume ratio of ignition zone/entire vessel in the 20 L vessel is larger than those in the 1000 L chamber. The temperature rise due to ignitors varies significantly in the 20 L vessel and not in the 1000 L vessel, which is an alternative reason.

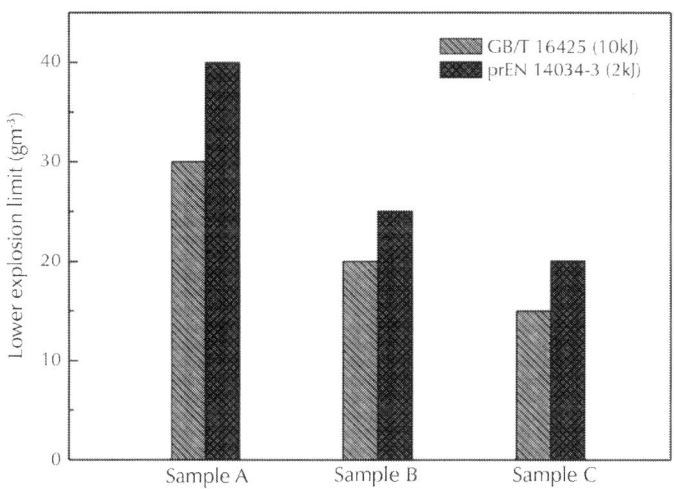

Figure 12: LEL for various particle size distributions, according to EN 14034-3 and GB/T 16425.

In practice, ignition source strength plays a key role in the LEL testing. Chawla et al., 1996 and Cashdollar and Chatrathi, 1993 and Going et al. (2000) found that 2500 J is the most suitable for LEL determination in the 20 L vessel. Another study of inerting of dusts found that 500–1000 J ignitors in the 20 L vessel most closely matches results in the 1000 L vessel (Dastidar & Amyotte, 2002a). Nevertheless, above-mentioned conclusions were drawn by testing carbonaceous and less volatile metal materials. Considering the overdriving phenomenon is not marked for magnesium dusts, it is essential to find out the most suitable ignition strength for LEL determination of magnesium dusts in the 20 L vessel.

Additional LEL testing was carried out for each sample with various ignition energies by applying the modified criterion PR ≥ 1.3 (Fig. 13). The stabilization of LEL is identified when ignition strength ranges from 2000 to 5000 J. This indicates that 2000–5000 J is the most appropriate energy region and realistic LEL date can be yielded over them. Furthermore, it is worth stressing that the stabilization is hardly seen again for carbonaceous materials like coal and oil shale dusts (Chawla et al., 1996). This is because the overdriving in the 20 L vessel is the greatest for materials with low volatility (Cashdollar and Chatrathi, 1993 and Going et al., 2000). Ignitors of 1000 J are not recommended for the LEL determination of magnesium dusts, because the corresponding results are away from the stabilization. Dust cloud with concentration reaching the flammability limit provides a bad situation for flame propagation due to worse inter-particle heat transfers and more heat losses. In the case of weak ignition strength, more dust load will be loaded to sustain the self-sustained flame propagation, which induces a higher LEL. Nevertheless, ignition energy of 10000 J might somewhat overdrive the poorly loaded dust clouds. The sufficient strong ignitors may preheat or burn the lean dust cloud within the flame of ignitors, even though the dust cloud could not sustained a self-sustained flame propagation (Myers, 2008). Consequently, GB/T 16425 is overly conservative for the LEL determination, whereas EN 14034-3 yields realistic LEL data. Moreover, the decrease of LEL with the rise of ignition energy somewhat demonstrates that LEL decreases with the rise of initial temperature (Eckhoff, 2003).

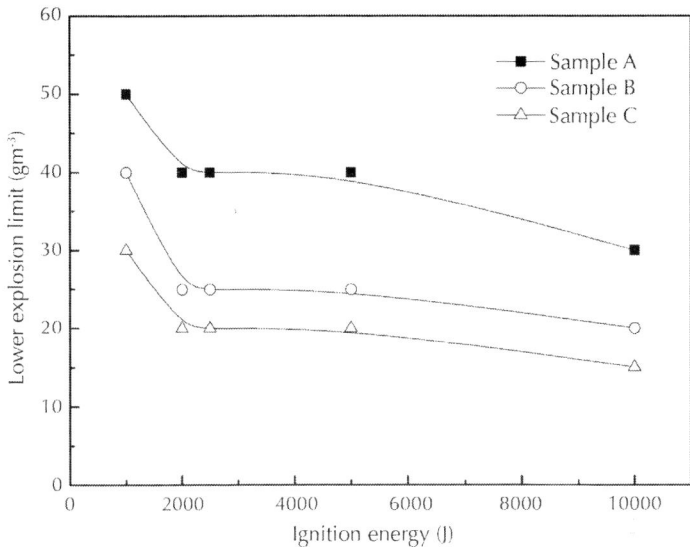

Figure 13: Evolutions of LEL with ignition energy for various particle sizes.

Influence of Initial Pressure

Two particle sizes were chosen to study the impact of initial pressure. Tests were carried out at four initial absolute pressures of 1.0, 1.2, 1.4 and 1.6 bar, using ignitors of 10000 J. In the present work, various initial pressures were obtained through different combinations between the pressure in the dust container and the vacuum in the vessel before dust dispersion.

Fig. 14a and b shows the influence of initial pressure on the explosion severity. It is clear that p_{max} and $(dp/dt)_{max}$ vary linearly with the rise of initial pressure. This behavior is expected because the augment of oxygen concentration leads to wider combustible concentration region for the same nominal dust concentration (Garcia-Torrent et al., 1998, Lazaro and Torrent, 2000 and Pilao et al., 2004) and the burning velocity is a strong function of oxygen concentration in fuel-rich mixtures (Goroshin et al., 1996). It is essential to point out that as particle size is raised, $(dp/dt)_{max}$ varies more significantly with the rise of initial pressure. This behavior is related to the surface Mg–O reactions. According to the heterogeneous mechanism, the overall reaction rate is controlled by the transport of oxygen to the particle surface. For larger particles, the disadvantage of specific surface area is significantly balanced by the rise of oxygen concentration in atmosphere.

Moreover, it is worth stressing that different evolutions of PR were obtained for various particle sizes (Fig. 14c). Results show a slight increment for sample A while the stabilization is identified for sample C. The increase of PR means that fewer heat losses are verified. The behavior is related to the increase of burning velocity associated with the rise of oxygen concentration. For larger particles, the reaction process will be more efficient and adiabatic under hyperbaric conditions (Pilao et al., 2004); therefore the amount of heat liberated during the overall combustion is raised. Alternatively, the increase of PR also means that the combustion of larger particles becomes more complete due to the rise of oxygen concentration.

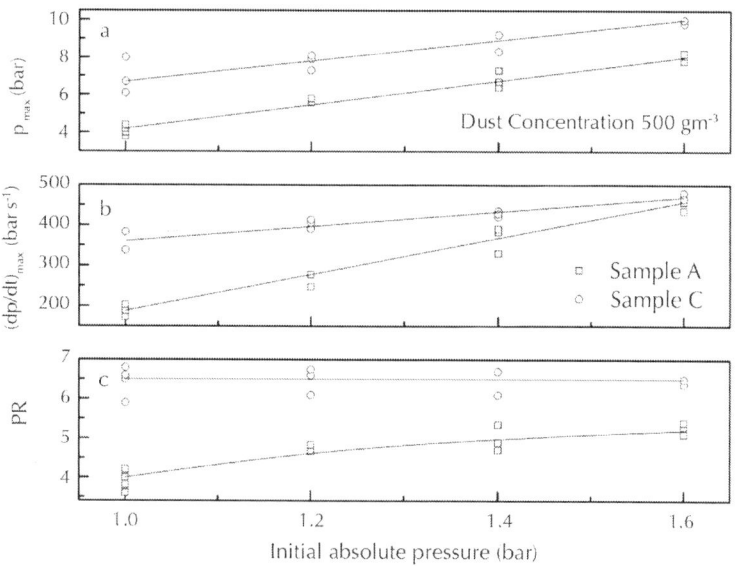

Figure 14: Evolutions of p_{max}, $(dp/dt)_{max}$ and PR with initial absolute pressure for various particle sizes.

Influence of Added Inertant

The experimental date, using KCl as interant, was shown in Fig. 15 for magnesium dust with four different levels of inertant. Two characteristics can be identified. Firstly, p_{max} decreases with the rise of mixing ratio of inertant. KCl represents significant inerting effectiveness up to a ratio of 60% for fuel–lean mixtures, and even can transform the fuel–inertant mixture into non-combustible. Secondly, as the magnesium dust concentration is raised, the inerting effectiveness becomes less significant. It demands a mixing ratio up to 80% to fully suppress the explosion when magnesium concentration is more than 500 g m^{-3}. This indicates that inerting requirement for fuel-rich dust cloud is raised. The experimental date, using $CaCO_3$ as interant, was shown in Fig. 16 for magnesium dust with two different particle sizes of inertant. Results show that the mitigation of explosion severity is associated with the rise of mixing ratio of $CaCO_3$. For 80% $CaCO_3$, the explosion is fully suppressed. Moreover, it is obvious that finer $CaCO_3$ powders represent better inerting effectiveness, and thereby it can be inferred that use of a coarser inertant raises the quantity of inertant required to inert. Moreover, the comparison of two inertants was illustrated in Fig. 17. Inerting effectiveness of $CaCO_3$ appears to be higher than KCl values on thermodynamics, whereas KCl powders represent higher inerting effectiveness upon kinetics.

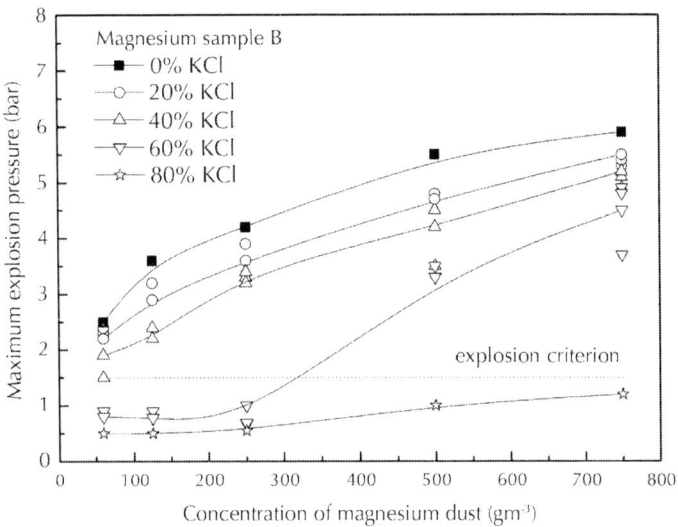

Figure 15: Evolutions of p_{max} with dust concentration for various mixing ratios of KCl, for magnesium particles with a median diameter of 22.4 μm.

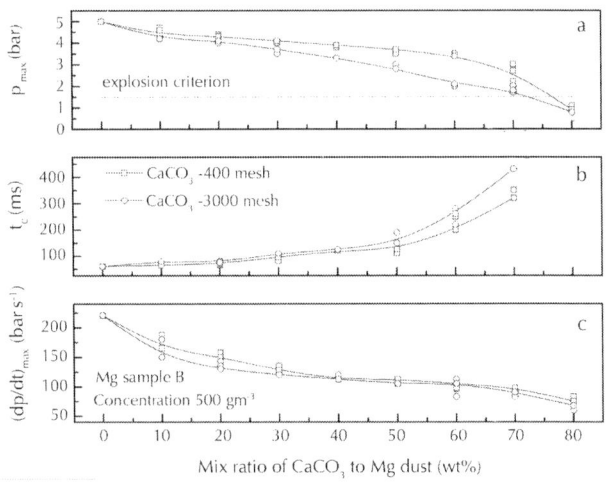

Figure 16: Evolutions of p_{max}, $(dp/dt)_{max}$ and t_c with mixing ratio of $CaCO_3$, for $CaCO_3$ powders with particle sizes of −400 and −3000 mesh, and magnesium dust (Sample B) with a concentration of 500 g m^{-3}.

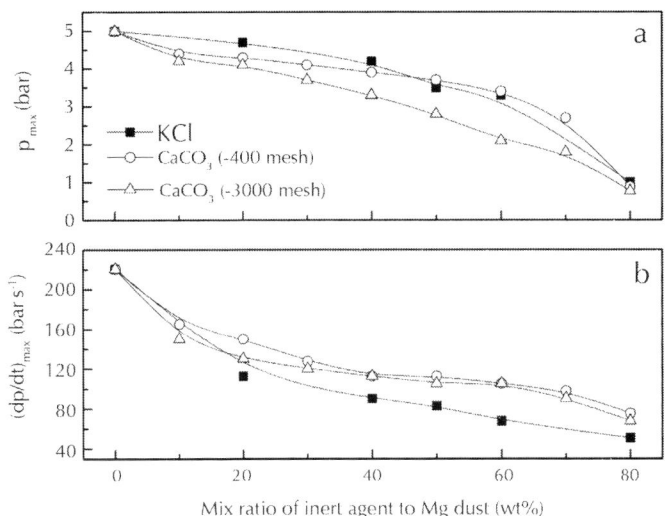

Figure 17: A comparison of inerting effectiveness between KCl and CaCO$_3$.

The inerting mechanisms are described as follows: on one hand, the inertant in the mixture lead to removal of the heat necessary for combustion (Amyotte, 2006) and the flame propagation would be prevented if the flame temperature is reduced below the limit value (Dastidar & Amyotte, 2004); on the other hand, the presence of inertant results in higher resistance in oxygen diffusion (Chatrathi & Going, 2000).

The heat balance of fuel–inertant mixture explosion addressed in literature (Chatrathi & Going, 2000) gives the idea that both the specific heat and the decomposition of inertant play a key role in inerting effectiveness, consistent with Abbasi and Abbasi (2007) and Amyotte's (2006) expressions. Powders of CaCO$_3$ represent higher inerting effectiveness on thermodynamics than KCl. On one hand, this is due to higher specific heat associated with CaCO$_3$. On the other hand, the thermal effect may also include the contribution from endothermic decomposition of CaCO$_3$; and CO$_2$ decomposed from CaCO$_3$ can effectively inhibit the magnesium dust flame propagation (Li, Yuan et al., 2009). It is well-established that a reduction in inertant particle diameter leads to enhanced inerting performance (Dastidar & Amyotte, 2002b). The reduction of inertant particle size leads to a corresponding increase in inertant particle surface area, which in turn leads to greater radiant heat absorption (Dastidar, Amyotte, & Pegg, 1997). Moreover, the possibility also exists that smaller inertant particles are able to decompose faster (Amyotte, Mintz, & Pegg, 1995). However, if the conclusion that finer size of a given inertant is always more effective at explosion inerting than coarser size was drawn; this statement may be limited, because ultra fine inertant may tend to agglomerate negating the intended benefit of a reduction in inertant particle size (Amyotte, 2006).

It is interesting to find that KCl powders represent higher inerting effectiveness upon kinetics while its lower inerting effectiveness on thermodynamics is verified. This is somewhat attributed to the particular efficacy of chloride in flame extinguishing (Horrocks & Price, 2001). Finally, it should be pointed out that neither KCl nor $CaCO_3$ is a very perfect inertant because both of them have to be added at a high ratio to suppress magnesium dust explosions completely, and thereby the more effective inertant needs further investigation.

CONCLUSIONS

The severe overpressure of magnesium dust explosions is emphasized in comparison with bituminous coal dusts. That magnesium dust is more of an explosion hazard than coal dusts is confirmed and quantified. As dust concentration is raised, the more efficient and adiabatic is the combustion during the explosion.

Trends of faster burning velocity and more efficient flame propagation are associated with smaller particles. The increase of explosion severity and explosibility with the diminishing of particle size gives the idea that finer particles represent greater hazards.

About 40–60% of the ignition energy is transferred to the atmosphere in the 20 L vessel. The ignition strength significantly affects the initial temperature in the 20 L vessel while the initial temperature appears to be stabilization in the 1000 L vessel. Different explosion behaviors of magnesium and bituminous coal dusts under various ignition energies are verified. The obtained results reveal that the overdriving phenomenon usually occurs for carbonaceous and less volatile metal materials is not notable for magnesium dusts.

The comparison between Chinese and European LEL determination procedures was accomplished. It seems that GB/T 16425 is overly conservative while EN 14034-3 yields realistic LEL data. The obtained results also reveal that 2000–5000 J is the most appropriate ignition energy to use in the LEL determination of magnesium dusts, using the 20 L vessel. Moreover, the experimental data is somewhat in agreement with the theoretic prediction.

The explosion severity increases in direct proportion with the rise of ignition pressure. Especially for larger particles, the combustion is more efficient and adiabatic under hyperbaric conditions.

The added KCl or $CaCO_3$ powders can mitigate the explosion severity and even transform fuel–inertant mixtures into non-combustible. Finer inertant shows better inerting effectiveness. Moreover, inerting effectiveness of $CaCO_3$ appears to be higher than KCl values on thermodynamics, whereas KCl represents higher inerting effectiveness upon kinetics.

ACKNOWLEDGEMENTS

This research was supported by 523 foundation of Sichuan university of China. Authors would like to acknowledge Zhu Li (Sichuan University) for particle size determination.

REFERENCES

1. Abbasi, T., & Abbasi, S. A. (2007). Dust explosions e cases, causes, consequences, and control. Journal of Hazardous Materials, 140(1e2), 7e44.

2. Amyotte, P. R. (2006). Solid inertants and their use in dust explosion prevention and mitigation. Journal of Loss Prevention in the Process Industries, 19(2e3), 161e173
3. Amyotte, P. R., Mintz, K. J., & Pegg, M. J. (1995). Effect of rock dust particle-size on suppression of coal-dust explosions. Process Safety and Environmental Protection, 73(B2), 89e100
4. Amyotte, P. R., Pegg, M. J., & Khan, F. I. (2009). Application of inherent safety principles to dust explosion prevention and mitigation. Process Safety and Environmental Protection, 87(1), 35e39
5. ASTM. (2007). Standard test method for pressure and rate of pressure rise for combustible dusts e E 1226. West Conshohocken: ASTM International
6. Cashdollar, K. L. (1994). Flammability of metals and other elemental dust clouds. Process Safety Progress, 13(3), 139e145
7. Cashdollar, K. L. (2000). Overview of dust explosibility characteristics. Journal of Loss Prevention in the Process Industries, 13(3e5), 183e199.
8. Cashdollar, K. L., & Chatrathi, K. (1993). Minimum explosible dust concentrations measured in 20 L and 1 m3 chambers. Combustion Science and Technology, 87(1e6), 157e171.
9. Cashdollar, K. L., & Zlochower, I. A. (2007). Explosion temperatures and pressures of metals and other elemental dust clouds. Journal of Loss Prevention in the Process Industries, 20(4e6), 337e348
10. CEN/TC305. (2004a). Determination of explosion characteristics of dust clouds e Part2: Determination of the maximum rate of explosion pressure rise (dp/dt)max of dust clouds. EN 14034e2. Brussels: European Committee for Standardization.
11. CEN/TC305. (2004b). Determination of explosion characteristics of dust clouds e Part1: Determination of the maximum explosion pressure pmax of dust clouds. EN 14303e1. Brussels: European Committee for Standardization.

12. CEN/TC305. (2004c). Determination of explosion characteristics of dust clouds e Part3: Determination of the lower explosion limit LEL of dust clouds. EN 14034e3. Brussels: European Committee for Standardization
13. Chatrathi, K., & Going, J. (2000). Dust deflagration extinction. Process Safety Progress, 19(3), 146e153.
14. Chawla, N., Amyotte, P. R., & Pegg, M. J. (1996). A comparison of experimental methods to determine the minimum explosible concentration of dusts. Journal of Loss Prevention in the Process Industries, 75(6), 654e658.
15. Dastidar, A., & Amyotte, P. (2002a). Determination of minimum inerting concentrations for combustible dusts in a laboratory-scale chamber. Process Safety and Environmental Protection, 80(B6), 289e299.
16. Dastidar, A. G., & Amyotte, P. R. (2002b). Explosibility boundaries for fly ash/ pulverized fuel mixtures. Journal of Hazardous Materials, 92(2), 115e126.
17. Dastidar, A. G., & Amyotte, P. R. (2004). Using calculated adiabatic flame temperatures to determine dust explosion inerting requirements. Process Safety and Environmental Protection, 82(B2), 142e155
18. Dastidar, A. G., Amyotte, P. R., & Pegg, M. J. (1997). Factors influencing the suppression of coal dust explosions. Fuel, 76(7), 663e670
19. Denkevits, A., & Dorofeev, S. (2006). Explosibility of fine graphite and tungsten dusts and their mixtures. Journal of Loss Prevention in the Process Industries, 19(2e3), 174e180.
20. Dreizin, E. L. (2000). Phase changes in metal combustion. Progress in Energy and Combustion Science, 26(1), 57e78.
21. Dreizin, E. L. (2003). Effect of phase changes on metal-particle combustion processes. Combustion Explosion and Shock Waves, 39(6), 681e693
22. Dreizin, E. L., Berman, C. H., & Vicenzi, E. P. (2000). Condensed-phase modifications in magnesium particle combustion in air. Combustion and Flame, 122(1e2), 30e42.

23. Dreizin, E. L., & Hoffmann, V. K. (1999). Constant pressure combustion of aerosol of coarse magnesium particles in microgravity. Combustion and Flame, 118(1e2), 262e280.
24. Dreizin, E. L., & Hoffmann, V. K. (2000). Experiments on magnesium aerosol combustion in microgravity. Combustion and Flame, 122(1e2), 20e29.
25. Dreizin, E. L., & Shoshin, Y. (2003). Particle combustion rates in premixed flames of polydisperse metal-air aerosols. Combustion and Flame, 133, 275e287.
26. Dufaud, O., Traore, M., Perrin, L., Chazelet, S., & Thomas, D. (2010). Experimental investigation and modelling of aluminum dusts explosions in the 20 L sphere. Journal of Loss Prevention in the Process Industries, 23(2), 226e236
27. Eckhoff, R. K. (2003). Dust explosions in the process industries (3rd ed.). Amsterdam: Gulf Professional Publishing
28. Evans, A. A. (1994). Deflagrations in spherical vessel: a comparison among four approximate burning velocity formulae. Combustion and Flame, 97, 429e434.
29. Gao, C., Li, H., Su, D., & Huang, W. X. (2010). Explosion characteristics of coal dust in a sealed vessel. Explosion and Shock Waves, 30(2), 164e168
30. Garcia-Torrent, J., Conde-Lazaro, E., Wilen, C., & Rautalin, A. (1998). Biomass dust explosibility at elevated initial pressures. Fuel, 77(9e10), 1093e1097.
31. Going, J. E., Chatrathi, K., & Cashdollar, K. L. (2000). Flammability limit measurements for dusts in 20 L and 1 m3 vessels. Journal of Loss Prevention in the Process Industries, 13(3e5), 209e219.
32. Goroshin, S., Fomenko, I., & Lee, J. H. S. (1996). Burning velocities in fuel-rich aluminum dust clouds. In 26th symposium (International) on combustion (pp. 1961e1967). The Combustion Institute.
33. Hertzberg, M., Cashdollar, K. L., Daniel, L. N., & Conti, R. S. (1982). Domains of flammability and thermal ignitability for pulverized coals and other dust: particle size dependences

and microscopic residue analyses. In 19th Symposium (International) on Combustion (pp. 1169e1180). The Combustion Institute.
34. Hertzberg, M., Cashdollar, K. L., & Lazzara, C. P. (1981). The limits of flammability of pulverized coals and other dusts. In 18th Symposium (International) on Combustion (pp. 717e729). The Combustion Institute.
35. Hertzberg, M., & Zlochower, I. A. (1990). Devolatilization rates and interparticle wave structures during the combustion of pulverized coals and polymethlmethacrylate. In 23rd Symposium (International) on Combustion (pp. 1247e1255). The Combustion Institute.
36. Hertzberg, M., Zlochower, I. A., & Cashdollar, K. L. (1992). Metal dust combustion: explosion limits, pressures, and temperatures. In 24th Symposium (International) on Combustion (pp. 1827e1835). The Combustion Institute.
37. Horrocks, A. R., & Price, D. (2001). Fire retardant materials. Boston: CRC Press
38. Huang, Y., Risha, G. A., Yang, V., & Yetter, R. A. (2009). Effect of particle size on combustion of aluminum particle dust in air. Combustion and Flame, 156(1), 5e13.
39. Kuai, N. S., Li, J. M., & Chen, Z. (2010). Study on the risk control of magnesium dust explosion based on inherent safety principle. Fire Science and Technology, 29(5), 369e372.
40. Lazaro, E. C., & Torrent, J. G. (2000). Experimental research on explosibility at high initial pressures of combustible dusts. Journal of Loss Prevention in the Process Industries, 13(3e5), 221e228.
41. Li, H., Gao, C., Su, D., & Huang, W. X. (2009). Experimental research on bituminous coal dust explosibility. Journal of Sichuan University (Engineering Science Edition), 41(6), 79e83.
42. Li, G., Yuan, C. M., Fu, Y., Zhong, Y. P., & Chen, B. Z. (2009). Inerting of magnesium dust cloud with Ar, N-2 and CO_2. Journal of Hazardous Materials, 170(1), 180e183.

43. Li, G., Yuan, C. M., Zhang, P. H., & Chen, B. Z. (2008). Experiment-based fire and explosion risk analysis for powdered magnesium production methods. Journal of Loss Prevention in the Process Industries, 21(4), 461e465.
44. Lutz, W. K. (1997). Advancing inherent safety into methodology. Process Safety Progress, 16(2), 86e88.
45. Matsuda, T., Yashima, M., Nifuku, M., & Enomoto, H. (2001). Some aspects in testing and assessment of metal dust explosions. Journal of Loss Prevention in the Process Industries, 14(6), 449e453.
46. MCI. (1996). Determination for minimum explosive concentration of dust cloud e GB/T 16425. Beijing: State Administration for Quality Supervision and Inspection and Quarantine of China.
47. Mintz, K. J., Bray, M. J., Zuliani, D. J., Amyotte, P. R., & Pegg, M. J. (1996). Inerting of fine metallic powders. Journal of Loss Prevention in the Process Industries, 9(1), 77e80
48. Mittal, M. (1997). Models for minimum explosible concentration of organic dust clouds handled in industries. Chemical Engineering & Technology, 20(7), 502e509
49. Myers, T. J. (2008). Reducing aluminum dust explosion hazards: case study of dust inerting in an aluminum buffing operation. Journal of Hazardous Materials, 159(1), 72e80.
50. Nifuku, M., Koyanaka, S., Ohya, H., Barre, C., Hatori, M., Fujiwara, S., et al. (2007). Ignitability characteristics of aluminium and magnesium dusts that are generated during the shredding of post-consumer wastes. Journal of Loss Prevention in the Process Industries, 20(4e6), 322e329.
51. Pilao, R., Ramalho, E., & Pinho, C. (2004). Influence of initial pressure on the explosibility of cork dust/air mixtures. Journal of Loss Prevention in the Process Industries, 17(1), 87e96
52. Pilao, R., Ramalho, E., & Pinho, C. (2006). Overall characterization of cork dust explosion. Journal of Hazardous Materials, 133(1e3), 183e195.

53. Proust, C., Accorsi, A., & Dupont, L. (2007). Measuring the violence of dust explosions with the "201 sphere" and with the standard "ISO 1 m3 vessel" e systematic comparison and analysis of the discrepancies. Journal of Loss Prevention in the Process Industries, 20(4e6), 599e606.
54. Rai, A., Park, K., Zhou, L., & Zachariah, M. R. (2006). Understanding the mechanism of aluminium nanoparticle oxidation. Combustion Theory and Modelling, 10(5), 843e859
55. Siwek, R. (1996). Determination of technical safety indices and factors influencing hazard evaluation of dusts. Journal of Loss Prevention in the Process Industries, 9(1), 21e31.
56. Tang, F. D., Goroshin, S., Higgins, A., & Lee, J. (2009). Flame propagation and quenching in iron dust clouds. Proceedings of the Combustion Institute, 32, 1905e1912
57. Zhang, C. G., Jiang, J. C., & Zheng, Z. Q. (2005). Study on the mode and prevention of dust explosion accident. China Safety Science Journal, 15(6), 73e76.
58. Zhen, G. P., & Leuckel, W. (1997). Effects of ignitors and turbulence on dust explosions. Journal of Loss Prevention in the Process Industries, 10(5e6), 317e324.

Chapter 5

Modelling the Effect of Particle Size on Dust Explosions

A. Di Benedetto[a], P. Russo[b], P. Amyotte[c], and N. Marchand[c]

[a]Istituto Ricerche Combustione, IRC-CNR, via Diocleziano, 328, 80125 Napoli, Italy

[b]Dipartimento di Ingegneria Chimica e Alimentare, Università di Salerno, via Ponte don Melillo, 84084 Fisciano (SA), Italy

[c]Department of Process Engineering and Applied Science, Dalhousie University, Halifax, NS, Canada B3J 2X4

ABSTRACT

The recent concept of inherent safety uses the properties of a material or process to eliminate or reduce the risk thus removing or minimizing the hazard at the source as opposed to accept the hazard and looking to mitigate the effects. In this framework the control of particle size in dust explosion prevention and mitigation is recognized as a major inherent safety methodology. Indeed, the increase of particle size may allow significant reduction of particle reaction rate eventually reducing the risk.

In this paper a novel model is developed to quantify the effect of particle size on dust reactivity in an explosion phenomenon. The model takes into account all of the steps involved in a dust explosion: internal and external heating, devolatilization reaction and volatiles combustion. Varying the dust size can establish different regimes depending on the values of the characteristic time of each step and of several dimensionless numbers (Damköhler number, Da; Biot number, Bi; thermal Thiele number, Th). Results from the model are reported in terms of the deflagration index (K_{St}) as a function of dust diameter in all regimes and at varying Da, Th and Bi. Comparison with experimental data from polyethylene explosion tests shows promising results. Finally, the results of the model are presented in the form of a dust explosion regime diagram, which is helpful to make a draft evaluation of the role of dust size on explosion behavior and severity.

INTRODUCTION

Many combustible dusts, if dispersed as a cloud in air and ignited, will allow a flame to propagate through the cloud in a manner similar to the propagation of flames in premixed fuel–oxidant gases (Proust, 2005). Such dusts include common foodstuffs such as sugar, flour and cocoa, synthetic materials such as plastics, chemicals and pharmaceuticals, metals such as aluminum and magnesium, and traditional fuels such as coal and wood.

In a previous paper we developed and validated a model for the evaluation of the thermo-kinetic parameters describing the maximum overpressure attained during an explosion (P_{max}), the deflagration index (K_{St}) and the burning velocity (S_l), which are usually required for the design of protection and mitigation systems against dust explosion hazards (Di Benedetto and Russo, 2007). The evaluation of such parameters was done by assuming that the pyrolysis/devolatilization step is very fast, leading to gas combustion controlling the dust explosion. This assumption is valid when the dust particle size is lower than a critical value (Eckhoff, 2003; Cashdollar et al., 1989); in this case the developed model allows the determination of the most conservative values of K_{St}, S_l and P_{max}.

When larger dust particles exist, other phenomena such as devolatilization and particle heating can control the explosion process. Under these conditions, the explosion severity is significantly reduced. An increase in particle size is beneficial when looking at reducing risk in the process industries. Indeed, in the framework of the application of the inherent safety principle of *moderation* (Amyotte et al., 2007), the avoidance of fine dust sizes allows the reduction of risk of dust explosions (Amyotte et al., 2008).

The qualitative and quantitative dependence of the violence of explosion on particle size is strongly affected by the interplay of phenomena controlling the combustion/heating of the solid material, such as devolatilization (where volatiles are given off by the particles or the particles are vapourized), gas-phase mixing and gas-phase combustion (Eckhoff, 2003), and by the dispersion of the dust particles in air. Each dust is characterized by a limiting particle size (asymptotic diameter of the order of 30μm), below which the combustion rate of the dust cloud ceases to increase, and by a particle size (of the order of 500μm) beyond which it may even become non-ignitable.

In the present paper we extend our previous model to take into account the effect of particle size on the deflagration index. The model considers all the steps which may be involved in the explosion of a dust: external and internal particle heating, pyrolysis/devolatilization reaction and volatiles combustion. Simulations were performed by varying the dust size, leading to different regimes depending on the value of the characteristic time of each step. Finally, the model results were compared with experimental data from polyethylene explosion tests.

MODEL DESCRIPTION

For organic dust clouds, devolatilization followed by gas-phase combustion has been proposed as the dominant mechanism of flame propagation (Eckhoff, 2003; Cashdollar et al., 1989; Hertzberg et al., 1988). In agreement with this mechanism, the model here developed proceeds through the following two steps in order to calculate the explosion parameters for a given dust:

- Calculation of the volatiles production rate as function of the dust chemical–physical properties and size;
- Calculation of the deflagration index of the volatiles explosion.

Both of these steps are described in the following sections.

Pyrolysis/Devolatilization

Models of pyrolysis of solids abound in the literature, but most involve one of two approaches. The simplest approach, often termed the ablation model, assumes that the solid decomposes to volatiles directly at a critical temperature (Andrews and Atthey, 1975). The second approach involves incorporating a kinetic mechanism for the degradation process, often inferred from thermogravimetric analyses, which allows the solid to decompose over a characteristic temperature range. This involves the use of one or more rate equations to model the rate of change in mass of a small sample as a function of remaining mass and temperature (Staggs, 2000; Wichman, 1986; Di Blasi, 1997).

The main consequence of employing finite-rate kinetics, rather than the critical-temperature approach of ablation models, is that pyrolysis occurs throughout the interior of the sample, rather than just at the exposed surface.

In the following, the second approach was applied. Pyrolysis kinetics is assumed to occur through two steps as typically occurs for thermoplastic polymers and cellulose (Di Blasi, 1999). Starting from the solid, S, the first step (k^0) describes the formation of an intermediate condensed-phase (molten phase), S^*, while the second step (k_1) represents the pyrolysis/devolatilization process with consequent formation of volatiles, V.

$$S \xrightarrow{k^0} S^* \xrightarrow{k_1} V \tag{1}$$

The model of volatiles production is developed under the following assumptions:
- One-dimensional, spherically symmetric system;
- Negligible resistance to mass transfer;
- No secondary reactions of volatile pyrolysis products;
- Local thermal equilibrium;
- Quasi-steady-state for the gas phase.

The model equations in dimensionless form are as follows:

Mass balance equations:

$$\frac{\partial M_s}{\partial t} = -k^0 M_s \tag{2}$$

$$\frac{\partial M_{s^*}}{\partial t} = k^0 M_s - k_1 M_{s^*} \tag{3}$$

$$\frac{\partial V}{\partial t} = k_1 M_{s^*} \tag{4}$$

where M_s is the actual mass of solid particle (S) divided by the initial mass of solid, M_{s^*} is the mass of intermediate condensed-phase species (S^*) divided by the initial mass of solid, V is the actual mass of volatiles (V) divided by the initial mass of solid.

Solid phase energy balance equation:

$$\rho c_p \frac{\partial T}{\partial t} = Q_r - c_V \frac{\partial (U\rho_v T)}{\partial x} + \frac{\partial}{\partial x}\left(\lambda \frac{\partial T}{\partial x}\right) \quad (5)$$

where ρ, c_p and λ are, respectively, density, specific heat and thermal conductivity of solid; ρ_v, c_v are, respectively, density and specific heat of volatiles, while Q_r is the heat produced by reaction: $Q_r = k^0 M_s H_s + k_1 M_{s^*} H_{s^*}$ where H_s and H_{s^*} are enthalpy of solid and of intermediate condensed-phase species, respectively. U is the outward velocity of the volatiles emitted from the shrinking solid particle.

The total volatiles fraction produced and leaving the dust surface at each time is evaluated from the integration of the volatiles mass balance equation along the particle diameter (d):

$$V_T = \int_0^d V\, dx \quad (6)$$

and, hence the volatiles production rate (VPR) at any time can be determined as follows:

$$VPR = \frac{dV_T}{dt} \quad (7)$$

The boundary conditions are where h_c is the heat transfer coefficient, ε is the emissivity, σ is the Stefan–Boltzmann constant and T_{ext} is the external temperature.

$$@x = 0 \text{ and } \forall t \quad \lambda \frac{\partial T}{\partial x} = 0 \quad (8)$$

$$@x = d \text{ and } \forall t \quad \lambda \frac{\partial T}{\partial x} = -h_c(T - T_{ext}) - \sigma\varepsilon(T^4 - T_{ext}^4) \quad (9)$$

The set of Eqs. (2), (3), (4), (5), (6) and (7) was solved with boundary conditions (8) and (9) by adopting the implicit finite difference scheme. The time integration was performed by means of the fourth order Runge–Kutta scheme with a time step equal to 10^{-6} s. For convenience, the model equations were solved in dimensionless form.

An important issue to consider in the modelling of thermal degradation of solid fuels is the determination of the decomposition regime so as to identify the process controlling step. The controlling step can be determined from the characteristic times of internal (t_c) and external (t_e, both convective and radiative) heat transfer and that of the chemical reaction (t_{pyro}).

According to Di Blasi (1999) the mechanism which controls the devolatilization process depends on the Biot number:

$$Bi = \frac{t_c}{t_e} = \frac{d(h_c \Delta T_i + \varepsilon \sigma \Delta T_i^4)}{\lambda \Delta T_i} \qquad (10)$$

where d is the dust diameter and cT_i is the temperature difference between particle and surrounding gases.

The Biot number, Bi, is a measure of the internal heat conduction time with respect to the external heat transfer time; hence, it can be defined as in Eq. (10). Two limit conditions can be observed, depending on the value of the Biot number:
- (case a) Bi1: the internal heat transfer rate is much faster than the external heat transfer rate and the thermal conversion process is dominated by the external heat transfer supply;
- (case b) Bi1: the internal heat transfer rate is much slower than the external heat transfer rate and thus internal heat transfer is the controlling mechanism.

Then the heat transfer times should be compared with those associated with chemical reaction.

For case a (Bi1), the characteristic times of external heat transfer (the slowest process) should be compared with the characteristic reaction time. This is usually done through the Damköhler number (Da), which can be expressed as where r_p is the pyrolysis reaction rate.

$$Da = \frac{t_e}{t_{pyro}} = \frac{r_p \Delta T_i c_p d}{h_c \Delta T_i + \varepsilon \sigma \Delta T_i^4} \quad (11)$$

For Bi1, two regimes can be observed:
- Regime I ($Bi1$ and $Da1$): when conversion occurs under the external heat transfer control;
- Regime II ($Bi1$ and $Da1$): when conversion occurs under the control of the pyrolysis chemical reaction.

For case b (Bi1), the characteristic time associated with internal heat transfer (the slowest process) should be again compared with the characteristic time of pyrolysis chemical reaction. This is usually done by means of the thermal Thiele number:

$$Th = \frac{t_c}{t_{pyro}} = \frac{r_p c_p d^2}{\lambda}$$
(12)

Thus, for Bi1, two regimes can be observed:
- Regime III (Bi 1 and Th 1): when conversion occurs under the pyrolysis chemical kinetic control;
- Regime IV (Bi 1 and Th 1): when conversion occurs under the control of internal heat transfer.

Explosion

Once the regime of the devolatilization process is identified, the step controlling the overall dust explosion phenomenon can be determined by comparing the characteristic time of the devolatilization controlling step with that relevant to the gas combustion (t_{comb}). At this level, the entire explosion phenomenon is modelled by calculating the laminar burning velocity of the volatilized dust particles dispersed in a sphere, at varying dust concentration, as described in our previous paper (Di Benedetto and Russo, 2007).

The assumption of this model is that the flame propagation is unaffected by the interaction between the dust particles.

To compare the step controlling the pyrolysis process to the combustion rate of the volatiles, we define a dimensionless number (Pc) given by the ratio of characteristic time of the pyrolysis/devolatilization reaction to that of volatiles combustion, as follows:

$$Pc = \frac{t_{pyro}}{t_{comb}} = \frac{\rho S_l}{r_p \delta_F} \qquad (13)$$

where δ_F is the flame thickness (typically, 1 mm) and S_l is the laminar burning velocity.

In the case of pyrolysis under chemical kinetic control (regime II or III), if Pc1 the pyrolysis reaction rate controls the overall explosion phenomenon; otherwise ($Pc1$), the volatiles combustion rate is the controlling step.

When regime I establishes and external heat transfer controls the conversion, the number $Da \cdot Pc$ should be evaluated:

$$Da \cdot Pc = \frac{t_e}{t_{comb}} = \frac{\rho c_p \Delta T_i d S_l}{(h_c \Delta T_i + \varepsilon \sigma \Delta T_i^4)\delta_F} \qquad 14)$$

When regime IV establishes, the internal heat transfer has to be compared to the combustion rate of volatiles trough the $Th \cdot Pc$ dimensionless number:

$$Th \cdot Pc = \frac{t_c}{t_{comb}} = \frac{\rho c_p d^2 S_l}{\lambda \delta_F} \qquad (15)$$

In consideration of the particle size, different regimes can occur: when the particle diameter is lower than a critical value (typical for any dust), the heating and pyrolysis/devolatilization steps are very fast (Pc;Da Pc and Th Pc 1) and then gas combustion is controlling the dust explosion. Conversely, in order to define the different regimes of dust pyrolysis the previous characteristic numbers (Bi, Da or Th) have to be evaluated.

We have previously demonstrated how to calculate the thermo-kinetic parameters which quantify the explosion behavior of dusts when the controlling step is the combustion of volatiles and then when Pc;Da Pc and Th Pc 1 (Di Benedetto and Russo, 2007). The deflagration index, maximum pressure and burning velocity calculated are valid for very low values of the dust size and then can be considered as asymptotic values.

In the present paper we evaluated these parameters by including the effect of the dust size. More precisely, we define χ (d) as follows:

$$\chi(d) = \frac{VPR_{max}}{VPR^0_{max}} \quad (16)$$

where VPR_{max} is the maximum volatiles production rate at a given dust diameter and VPR0 max is the maximum volatiles production rate at dust diameter approaching to zero.

RESULTS

Simulations were performed in each regime by varying the characteristic dimensionless numbers to obtain the values of χ factor in all regimes.

Simulations were done assuming a constant temperature of gas surrounding the dust particles of 2600 K and a gas velocity equal to 2 m/s, which corresponds to a fully developed turbulent regime typical of the standard 1 m³ and 20 l spheres.

Results of these simulations are reported in Fig. 1 in terms of the χ factor as a function of the dust diameter as obtained in each regime.

It appears that the most important variations with dust size are attained at regime IV where internal heat transfer controls the volatiles production. Also in regime I in which external heat transfer is the controlling step, the χ factor reaches about 25% the value when the diameter is about 500 μm.

Smoother variations occur in regimes II and III where the volatilization reaction rate is the controlling mechanism.

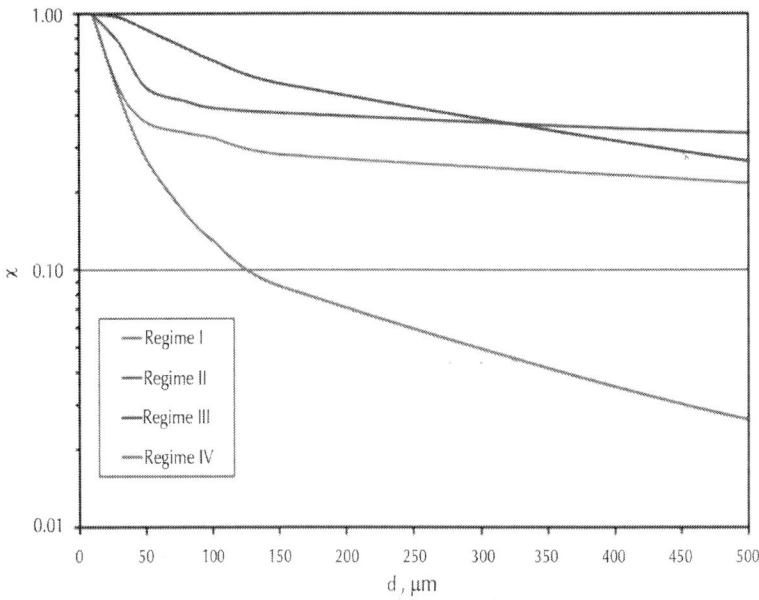

Figure 1: χ factor at varying dust size in different regimes as calculated by the model.

From Fig. 1 it appears that at very high values of the dust diameter (order of 500 μm), in regimes I, II and III, the χ factor is still high (~0.22–0.34), suggesting that in these cases the pyrolysis process does not significantly reduce the severity of the explosion phenomenon.

On the other hand, in regime IV, it is found that the χ factor is of the order of 0.03, meaning that the deflagration index is reduced by two orders of magnitude. In these conditions, moderation is due to the limit of the pyrolysis process.

In Fig. 2 the dust spatial/temporal temperature profiles are shown at different times as a function of x/d (where x is the thickness of solid and d is the particle diameter) and at d=10, 50, 100 and 500 μm.

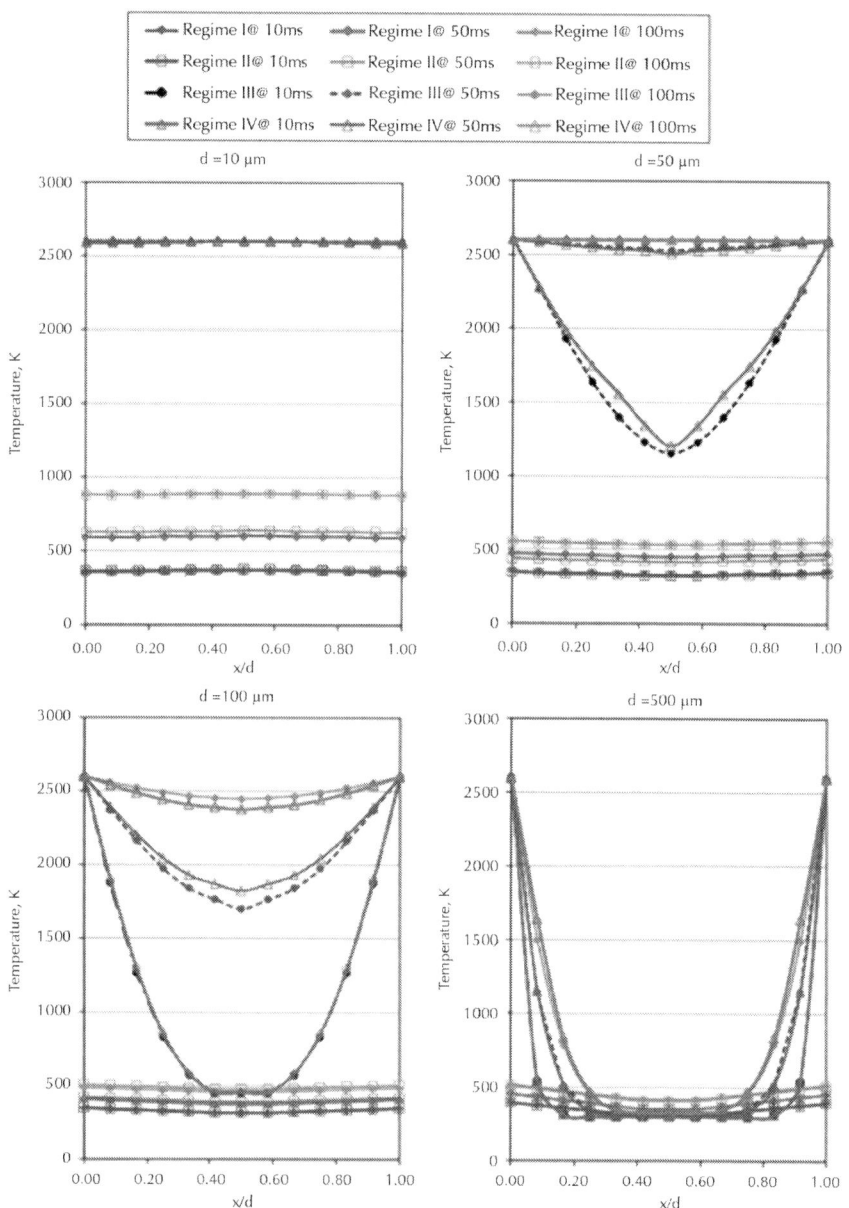

Figure 2: Temporal/radial temperature profiles at varying dust size in different regimes.

Depending on the regime at which the pyrolysis reaction proceeds, different radial temperature profiles, temperature levels and rate of temperature rise establish.

- *Regime I pyrolysis under external heat transfer control*: The temperature profile along the dust particle is flat at any time. The temperature level is limited by the external heat transfer and it increases faster at low values of the dust size.
- *Regime II pyrolysis under kinetic control* (but Bi51): The temperature profiles are similar to those observed in regime I. The temperature level is controlled by the occurrence of the reaction and eventually by the external heat transfer; it increases faster at low values of the dust size.
- *Regime III pyrolysis under kinetic control* (but Bi5 $\ll$ 1): At small dust size (10 µm) the radial temperature profile is flat, while on increasing the dust size a radial gradient is observed (from about 1100 K in the particle center to 2600 K at the surface, d=50 µm). In this regime the temperature level reaches high values similar to the gas temperature (T=2600 K).
- *Regime IV pyrolysis under internal heat transfer control*: In these conditions the entire process is controlled by the temperature level inside the dust particle. On increasing the dust diameter the temperature strongly varies from the surface to the center.

As expected in regimes I and IV, temperature profiles inside the dust particle are primary related to the heat transfer controlling mechanism (internal or external) and hence to the *Bi* value.

In the case of regimes II and III, both controlled by the pyrolysis chemical reaction, the temperature profiles inside the particle are determined by the heat transfer mechanism up to the temperature value to which the pyrolysis reaction starts to occur. When pyrolysis occurs, since the reaction is endothermic and with high activation energy, a lower temperature is observed in the particle in the case of regimes II and III with respect to regimes I and IV, respectively. The differences in temperature between regimes III and IV clearly appear in Fig. 2 for particles of 50 µm at 10 ms and 100 µm at 50–100 ms.

To study the effect of these temperature profiles on the volatilization reaction, the degree of solid conversion as a function of time was plotted for the dust sizes investigated for each regime (Fig. 3). In all regimes, the larger the dust particle, the slower the volatilization process.

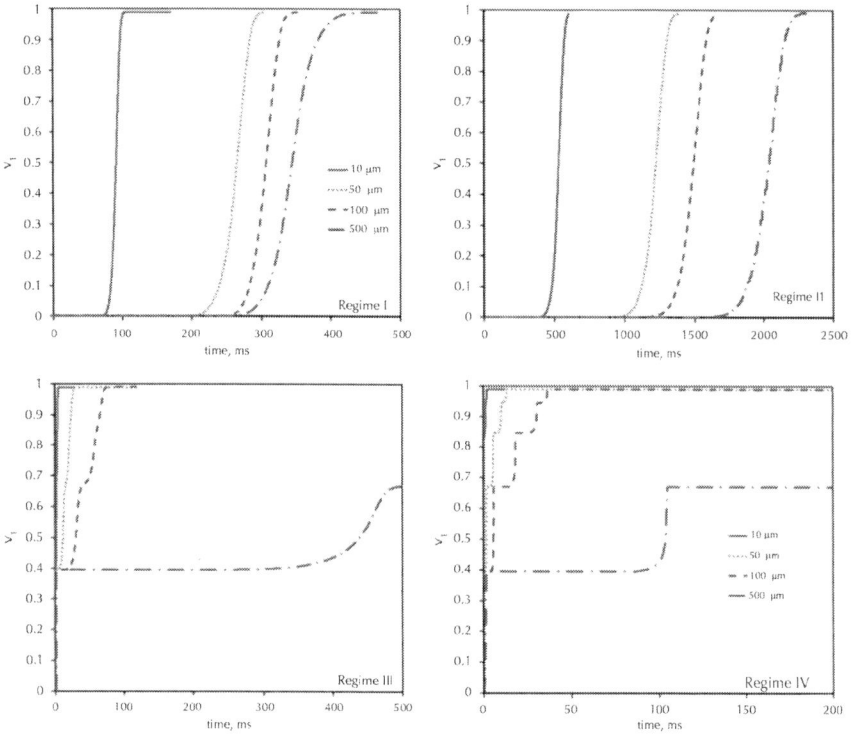

Figure 3: Conversion to volatiles at varying dust size in each regime.

In regimes I and II the temperature profile is not controlled by the internal heat conduction, and the temperature is almost constant inside the dust particle ($Bi \ll 1$). In regime I volatilization is controlled by the temperature level established in the particle and, more precisely, by the external heat transfer while the volatilization reaction is much faster. In regime II the kinetics of volatilization are slower than the heat transfer. Therefore, volatiles production is faster in regime I.

When the temperature profile is controlled by the internal heat transfer along the particle (Bi1); regimes III and IV), the volatilization process occurs much faster in the case of regime IV in which the kinetics of volatilization are fast compared to the internal heat transfer.

In particular, in regime III for a particle size of 50 μm at 10 ms, only a small fraction of solid (about 47%) is converted to volatiles while the conversion is almost complete (88%) in regime IV. Differences in solid conversion to volatiles were observed also for larger particle sizes (100 and 500 μm) at a longer time.

The time for volatiles production must be compared to the combustion time of the volatiles. From this comparison, the step controlling the dust explosion process will be identified.

Effect of Particle Size on Polyethylene Explosion

In a previous paper (Amyotte et al., 2008) we studied the explosion behavior of polyethylene by varying the dust size. Peak pressure and maximum rate of pressure rise at varying dust sizes (from 28 up to 916 μm) and concentrations were measured. In the current work, the polyethylene pyrolysis and explosion processes were simulated to enable comparison with the experimental results.

The pyrolysis kinetic and thermodynamic parameters used for polyethylene are reported in Table 1.

Table 1: Physical and kinetic parameters for simulation of polyethylene dust

Parameter	Value	Reference
ρ (kg/m³)	920	Rabinovitch (1965)
c_p (J/kg K)	2300	Staggs (2000)

λ (W/m K)	0.335	Staggs (2000)
A_0 (1/s)	6.19×10⁴	Rabinovitch (1965)
E_0 (J/mol)	71 000	Rabinovitch (1965)
A_1 (1/s)	5.20×10¹¹	Tewarson and Pion (1976)
E_1 (J/mol)	201 000	Tewarson and Pion (1976)
ΔH_0 (J/kg)	0	Proust (2005)
ΔH_1 (J/kg)	960 000	Wichman (1986)

Gas temperature was assumed to be equal to the adiabatic value at stoichiometric conditions and constant volume (T=2600 K), and the gas velocity was taken as 2 m/s. The initial dust temperature was assumed to be 300 K.

In Table 2 the Bi, Da and Th numbers are given as calculated for the polyethylene dust at various diameters.

Table 2: Bi, Da and Th dimensionless numbers and pyrolysis regimes of polyethylene dust

d (μm)	Bi	Da	Th	Pyrolysis regime
28	0.26	0.02	5.5×10⁻³	II
49	0.32	0.05	1.7×10⁻²	II
103	0.45	0.17	7.4×10⁻²	II
171	0.59	0.34	2.0×10⁻¹	–
276	0.80	0.67	5.3×10⁻¹	–
916	1.92	3.05	5.85	IV

In the range of dust size investigated (28–916 μm), external heat transfer prevails with respect to the internal heat transfer at low values of the dust diameter (Bi<0.5 at 28–103 μm), while at higher values of d the internal heat transfer is the controlling mechanism. (Bi>1 at 916 μm).

Since Da1 at 28–103 μm and Th1 at 916 μm, the polyethylene pyrolysis is controlled by the intrinsic kinetics (regime II) at low values of the dust size and by internal heat transfer (regime IV) at high values of the dust size.

The Pc, $Da \cdot Pc$ and $Th \cdot Pc$ dimensionless numbers defined in Eqs. (13), (14) and (15) were evaluated for polyethylene at the conditions reported in Table 2. These values calculated for particles assuming a laminar burning velocity (S_l) of 1 m/s and a flame thickness (δ_F) of 1 mm, are reported in Table 3.

Table 3: Pc, $Da \cdot Pc$ and $Th \cdot Pc$ dimensionless numbers for polyethylene dust explosion

d (μm)	Pc	Da·Pc	Th·Pc	Pyrolysis regime
28	906	19	5	II
49	906	47	15	II
103	906	149	67	II
171	906	312	185	–
276	906	603	481	–
916	906	2760	5300	IV

The results in Table 3 for $Pc = t_{pyro}/t_{comb}$ 1 mean that whatever the dust diameter value may be, the pyrolysis reaction rate is slower than the gas combustion rate. As a consequence it is possible to say that the pyrolysis is always controlling the explosion phenomenon, at least when $S_l = 1$ m/s. Similarly, at low values of laminar burning velocity, in the range 0.17–1 m/s as calculated by the model for polyethylene at different dust concentrations in a 20 l sphere, the values of Pc1 (Table 4) indicate that the pyrolysis is the controlling mechanism.

Table 4: Pc values at different S_l for polyethylene dust

S_l (m/s)	Pc
0.17	154
0.2	181

0.4	362
0.6	543
0.8	725
1.0	906

In the case of pyrolysis regime II (d=28–103 µm) with the pyrolysis reaction being the controlling step and $Pc \ll 1$, the pyrolysis reaction controls the overall explosion phenomenon.

For pyrolysis regime IV under internal heat transfer control (d=916 µm), with $Th \cdot Pc = t_c/t_{comb} \ll 1$, the internal heat exchange by conduction is the controlling step of the explosion phenomenon.

Finally, for intermediate sizes (171–276 µm), the values of $Th \cdot Pc$ and $Da \cdot Pc > 1$ indicate that internal and external heat exchange also play a role.

Simulations were run for polyethylene by varying the dust size and solving Eqs. (2), (3), (4), (5), (6) and (7) with boundary conditions (Eqs. (8) and (9)).

The model is used to calculate the χ parameter as defined in Eq. (16) and, hence, the ratio of the deflagration index with respect to the asymptotic value ($d \to 0$). The χ values at different polyethylene particle sizes are reported in Table 5. The asymptotic value of the deflagration index was taken as equal to $K_{St}^0 = 209$ barm/s, as reported in our previous paper (Di Benedetto and Russo, 2007).

Table 5: χ values for polyethylene dust

d (µm)	χ
28	0.65
49	0.57
103	0.41
171	0.19
276	0.16
916	0.02

Results of these simulations are here compared to our previous experimental data obtained in a standard 20 l sphere.

In Fig. 4 we plotted the model values of χ against the ratio between the experimental data of K_{St} and the asymptotic value of the deflagration index, K_{St}^0 ($K_{S:}/K_{S:}^0$).

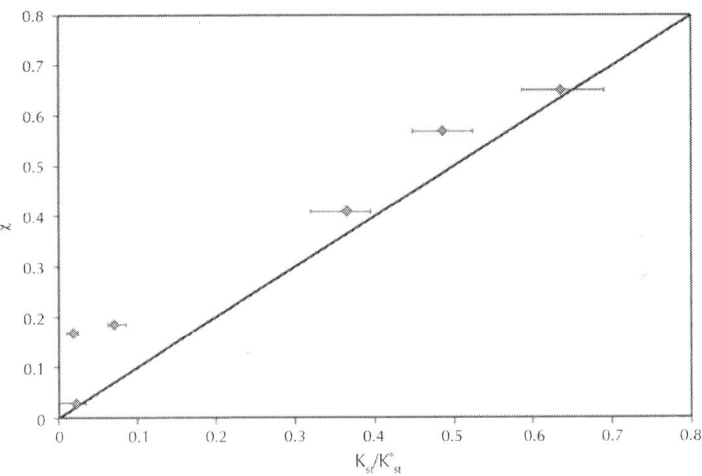

Figure 4: Comparison between the model values of χ and the experimental data of $K_{S:}/K_{St}^0$ of polyethylene ($K_{St}^0 = 209$ barm/s).

Where K_{St}^0 is the deflagration index calculated by assuming that the controlling step is the volatiles explosion at dust diameter approaching to zero ($d \approx 0$) and χ is previously defined (Eq. (16)).

The values of K_{St}^0 may be obtained as described in our previous paper (Di Benedetto and Russo, 2007).

The comparison between the experimental and model values suggest that a link between χ and the deflagration index exists, which is given by the following equation:

$$K_{St}(d) = K_{St}^0 \cdot \chi(d) \tag{17}$$

From Eq. (17) it comes out that once the asymptotic value of the deflagration index (K_{St}) is known, the effect of particle size on the deflagration index can be evaluated according to the pyrolysis model presented in the current work.

In Fig. 5 the comparison between the experimental and model data of K_{St} for polyethylene is shown.

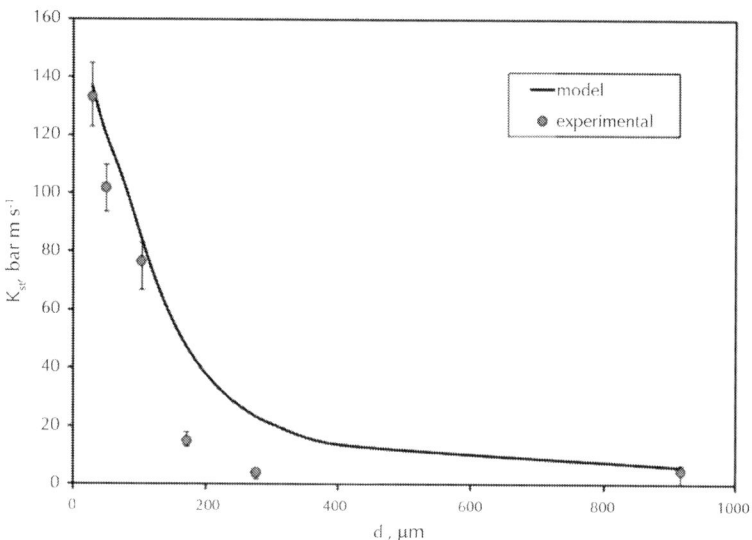

Figure 5: Deflagration index of polyethylene as function of dust diameter as obtained by the model and by experiments ($K_{St}^0 = 209$ barm/s).

It is found that at $d=103$ μm there occurs the passage from regime II, in which pyrolysis chemical reaction controls, to regime IV ($d>103$ μm) in which the internal heat transfer controls.

DISCUSSION

The model presented here is based on the assumption that dust pyrolysis occurs according to the shrinking core particle model. As a consequence, the current results apply to dust particles which react at constant volume. For all these particles, whatever their composition, the results reported in Fig. 1 may be easily used for evaluating the degree of mitigation which can be obtained by varying the dust size.

Once the deflagration index at diameter approaching zero is known, and the pyrolysis controlling mechanism has been identified, the χ index may be evaluated at each diameter from Fig. 1, thus evaluating the hazard moderation arising from coarse particles.

As an example, one may evaluate the χ factor of polyethylene. In Table 2 it is presented that the pyrolysis process of polyethylene occurs in regime II at diameters lower or equal to 103 μm. From Fig. 1 we may find χ=0.75, 0.52 and 0.43 at 28, 49 and 103 μm, respectively. Starting form K_{St}^0 = 209 barm/s, we obtain K_{St}=155, 109 and 90 bar/m s. Comparing these values with those found from our simulations of the entire process (Fig. 5), excellent agreement is obtained.

CONCLUSIONS

A model for dust explosion is here developed to quantify the effect of dust size on the dust hazard. This model couples all the steps involved in dust explosion: devolatilization, internal and external heat transfer and volatiles combustion.

Depending on the characteristic times of these processes, different explosion regimes establish giving rise to different trends of the violence of explosion, i.e. the deflagration index. Experimental values of the deflagration index for polyethylene at various dust particle sizes were successfully compared with the model values.

In the framework of inherent safety applied to the prevention and mitigation of dust explosion this model may help in quantifying the limiting dust size for process hazard moderation.

REFERENCES

1. Amyotte, P., Marchand, N., Di Benedetto, A., Russo, P., 2008. In: Proceedings of Seventh International Symposium Hazard Prevention Mitigation of Industrial Explosion, St. Petersburgh, vol. 3, pp. 103–113.
2. Amyotte, P.R., Pegg, M.J., Khan, F.I., 2007. Application of inherent safety principles to dust explosion prevention and mitigation. In: Proceedings of 12th International Symposium on Loss Prevention and Safety Promotion in the Process Industries, IChemE Symposium Series no. 153, Edinburgh, UK, May 22–24.
3. Andrews, G., Atthey, D.R., 1975. Analytical and numerical techniques for ablation problems. In: Ockendon, J. (Ed.), Moving Boundary Problems in Heat Flow and Diffusion. Clarendon Press, Oxford.
4. Cashdollar, K.L., Hertzberg, M., Zlochower, I.A., 1989. Effect of volatility on dust flammability limits for coals, gilsonite, and polyethylene. In: Proceedings of the 22nd Symposium (International) on Combustion, the Combustion Institute, Pittsburgh, pp. 1757–1765.
5. Di Benedetto, A., Russo, P., 2007. Thermo-kinetic modelling of dust explosions. Journal of Loss Prevention in the Process Industries 20, 303–309.
6. Di Blasi, C., 1997. Linear pyrolysis of cellulosic and plastic waste. Journal of Analytical and Applied Pyrolysis 40–41, 463–479.
7. Di Blasi, C., 1999. Transition between regimes in the degradation of thermoplastic polymers. Polymer Degradation and Stability 64, 359–367.
8. Eckhoff, R.K., 2003. Dust Explosions in the Process Industries. Gulf Professional Publishing.
9. Hertzberg, M., Zlochower, I.A., Cashdollar, K.L., 1988. Volatility model for coal dust flame propagation and extinguishment. In: Proceedings of the 21st Symposium

(International) on Combustion, the Combustion Institute, Pittsburgh, pp. 325–333.
10. Proust, C., 2005. A few fundamental aspects about ignition and flame propagation in dust clouds. Journal of Loss Prevention in the Process Industries 19, 104–120.
11. Rabinovitch, B., 1965. Regression rate and the kinetics of polymer degradation. In: Proceedings of 10th International Symposium on Combustion, the Combustion Institute, Pittsburgh, pp. 1395–1397.
12. Staggs, J.E.J., 2000. A simple model of polymer pyrolysis including transport of volatiles. Fire Safety Journal 34, 269–280.
13. Tewarson, A., Pion, R.F., 1976. Flammability of plastics. Combustion and Flame 26, 85–103.
14. Wichman, I.S., 1986. A model describing the steady-state gasification of bubbleforming thermoplastics in response to an incident heat flux. Combustion and Flame 63, 217–229.

Chapter **6**

A Reaction Engineering Approach to Modeling Dust Explosions

Vimlesh Kumar Bind[a,b], Shantanu Roy[a], and Chitra Rajagopal[b]

[a]Department of Chemical Engineering, Indian Institute of Technology – Delhi, Hauz Khas, New Delhi 110 016, India
[b]Center for Fire, Explosive and Environment Safety, Defence Research and Development Organization, Timarpur, Delhi 110 054, India

ABSTRACT

Dust explosions are a major hazard frequently encountered in vital sectors like food, energy, defense (propellants and explosives) and pharmaceuticals. These explosions emanate from rapid combustion of clouds of suspended fine particles in the micron range. Quantitative estimation of dust explosion propagation is crucial to their mitigation, and for providing estimates of physical quantities that determine the explosion behavior for design of safety systems. In this contribution, a multi-scale reaction engineering approach for modeling dust explosions has been presented. In the model, two scales are considered. At the particle scale, a detailed model involving various transport steps is written and solved for a variety of different boundary conditions. This model is used to build an effective reaction rate term which is incorporated in the dust cloud-scale CFD model through an appropriate source term. The CFD model for the effective dust cloud mixture is then executed to model the propagation of the dust explosion. Validation of the developed CFD model has been carried out for two different kinds of dust particles, aluminum (metallic dust) and starch (organic dust), for experimental data published in literature. Finally, a case study has been presented which shows the applicability of present approach in modeling real situations. It is demonstrated that using the same physics and estimated kinetic parameters from particle scale model or smaller scale experiments, fairly satisfactory prediction for dust explosions in real geometries could be obtained.

INTRODUCTION

Dust explosions involve rapid combustion of fine, combustible dust particles (generally in the micron size range). Following this rapid combustion, typically the pressure of product gases, in the enclosure containing the dust-oxidizer mixture, increases many folds within a fraction of second. This causes the "explosion" and can potentially cause collapse of the enclosure itself, or lead to secondary dust explosions which could be equally damaging to life and property even at some distance from the original point of the explosion. Historically the first recorded incidence of a dust explosion was at an Italian flourmill in 1785, although it was almost certainly not the first to occur [1]. These hazards are frequently

A Reaction Engineering Approach to Modeling Dust Explosions 169

encountered in vital industries and sectors of a country including agriculture, food processing, coal mining, defense (involving solid, granular explosives and propellants), plastic, wood and other materials processing, and pharmaceuticals. In these industries, various unit operations like storage, grinding, transportation and pneumatic conveying are susceptible to dust explosion hazards. Several excellent texts are now available which provide excellent reviews of this important but relatively less studied safety hazard (for example, see the works of Eckhoff [1], Field [2], Palmer [3] and Field [4]). However, most of these books deal with dust explosion as a safety and hazard issue, discuss ways of mitigating and preventing these, but few treat the underlying physics in great detail.

Proper understanding of dust explosion phenomena is necessary for design of any preventive or protective system to prevent such hazards causing accidents. Since the explosions result in rapid pressure build-up and heat localization, it is crucial to estimate the (primarily convective) transport rates with accuracy, so that a protection system could be designed and safety protocols may be established. This is of particular relevance since experiments are difficult to perform in these situations, and investigations leading to quantitative estimates can only be performed in well-designed, small-scale laboratory units, such as the classical Siwek 20L apparatus (e.g. [5]). Dust explosion experiments in actual process plants are rarely performed, due to the prohibitively high cost and risk involved ([6] is a notable exception).

Extensive work on modeling dust explosions using CFD [7], [8] and [9] in conjunction with input from laboratory scale experiments [10], [11] and [12] has revealed that much more work is needed on the basic physics and chemistry. Ogle [13] was perhaps the first one to emphasize the need of dust explosion prediction using material properties of the explosible mixture. Ogle [13] has developed a model for simulation of aluminum/air explosions in closed vessel, assuming spatially uniform pressure at any instant. Relation between particle scale surface burning rate and volumetric reaction rate at cloud scale was shown. However in this work particle scale modeling was not attempted and surface reaction rate was assumed as rate controlling mechanism for aluminum particle combustion. Further, 20 L spherical vessel was assumed as batch reactor for dust explosion reaction modeling. These simplified assumptions (may be due to lack of computational power) resulted in disagreement between Ogle's [13] theoretical and experimental work. Nevertheless this work suggested an important methodology based on material properties of combustible mixture and cloud scale hydrodynamics for dust explosion prediction.

This contribution was motivated by the fact that proper understanding of dust explosion phenomenon is necessary at "particle scale" as well as at "cloud scale". "Particle scale" deals with mechanism of ignition, gas–solid non-catalytic reaction, all at the size or length scale of the dust particles (which is usually a few microns). The transport phenomena at the "cloud scale" deals with flame and pressure propagation, which is actually in a two-phase gas–solids mixture (referred to as the "cloud"), even though for many purposes it may behave like a pure gas cloud. This overall vision of the dust explosion mechanism is reflected schematically in Fig. 1. This understanding and distinction of scale-wise transport should lead us to the formulation of a mathematical model to relate factors, parameters and design variables. When coupled with computational fluid dynamics (such as the work of Skjold et al. [8]), it is possible to translate the particle-scale and cloud-scale physics to complex geometries.

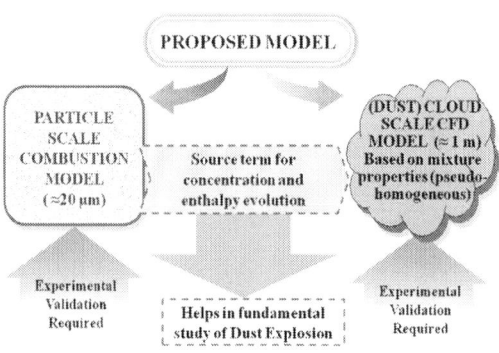

Figure 1: Proposed multi-scale approach for dust explosion modeling.

In an earlier contribution [14], a CFD model has been presented in which the dust-air mixture is viewed as an effective gas, and the explosion was driven by an effective source term resulting from particle-scale combustion. In this contribution, we present the combined model (involving particle-scale combustion and cloud-scale transport), show validation of this model and extend it to model a case study. Even with the assumption of quasi-homogeneous dust cloud, the simulation for dust explosion in industrial size of vessel ~10.3 m³ presented in proposed manuscript takes considerable time of the order of month with IBM Server: 64 bit Operating System, Processor Intel(R) Xenon (R) CPU, 2.27 GHz, 4.0 GB Memory. This is due to the fact that the time step used in proposed simulation is ~10^{-5} s where as total time of combustion in the vessel is of the order of ~1.0 s. The main advantage of proposed model compared with existing models is that it is based on material properties of fuel and oxidizer at particle scale and at cloud scale CFD could be applied which will make it more generalized and with lesser assumptions compared to existing models. This will also lead to fundamental understanding of dust explosion phenomenon. This model is versatile and could be applied for any geometry.

REPRESENTING PARTICLE SCALE PHENOMENA IN DUST CLOUD CFD

From our presentation of Fig. 1, it is clear that we intend to develop an "effective" description of particle scale phenomena and then connect it to the cloud scale CFD model. Descriptions at particle scale can be analytical, numerical or empirical (through correlations). Either of these descriptions are sensitive to the nature of the dust particles, the oxidizers, and most importantly, the average inter-particle distance in the cloud. This factor actually determines the kind of combustion the dust particles would exhibit, and hence what role it would have in determining the overall explosion behavior.

If these particles are separated by a large distance, single particle combustion will define the combustion behavior of the cloud. However if the concentration of dust particle is high and particles are in the vicinity, then the so-called *Group Combustion* or *Cloud Combustion* modes is observed [15]. Therefore, the average inter-particle distance is an important factor in defining the cloud scale dust explosion phenomenon.

According to Eckhoff [1], typical dust concentration for which explosion can occur ranges from 10 to 1000 g/m³. Density of most of the dust particles are of the order of 1000 kg/m³. Therefore, the typical value of particle volume fraction is of the order of $\phi_p = 10^{-5} - 10^{-3}$. The average inter-particle distance between particles as a function of volume fraction, in an assumed "particle-in-box" configuration shown in Fig. 2, is given by [16]:

$$\phi_p = \frac{(\pi/6)d_p^3}{L_p^3} \iff \left(\frac{L_p}{d_p}\right) = (\pi/6\phi_p)^{1/3} \qquad (1)$$

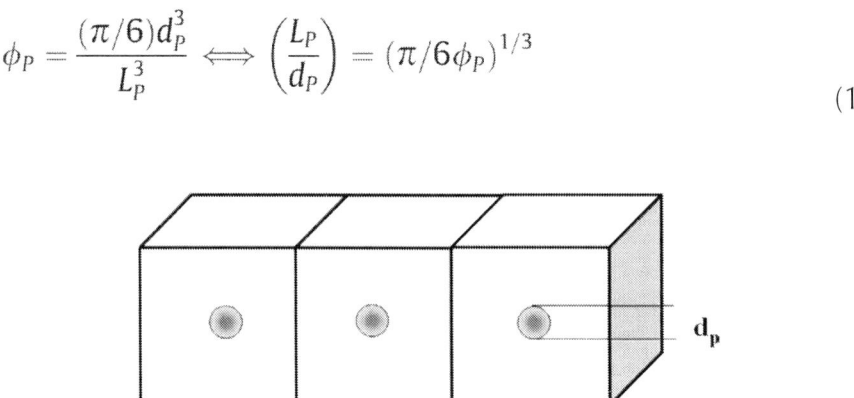

Figure 2: Assumed "particle-in-box" configuration of dust (spherical) cloud in air.

The relationship is shown in Fig. 3 as a ready reference. Naturally, the actual distribution of particles in a dust cloud need not follow the "particle-in-box" configuration (Fig. 2), but nevertheless that configuration offers us a reference point to determine what kind of combustion the cloud is experiencing. According to Annamalai et al. [15], the individual particle combustion will take place if the inter-particle distance (L_p) is at-least twice the flame radius of the particle. In this case, the group combustion number $d_p/L_p \ll 1.0$.

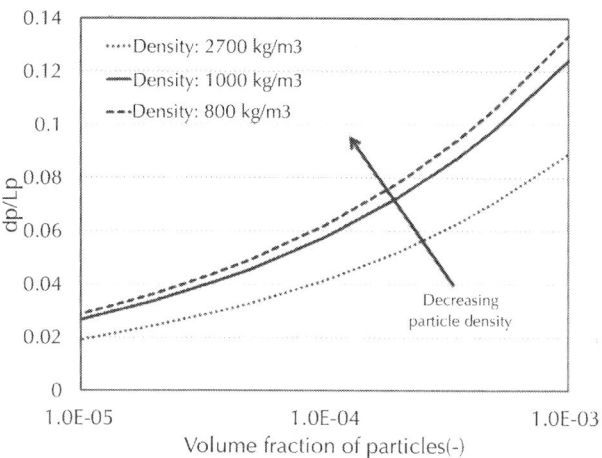

Figure 3: Inter-particle distance as a function of volume fraction of particle phase.

Metallic Particles

Consider the dust explosion of aluminum-air mixture with average concentration of aluminum of 0.35 kg/m³, which corresponds to a concentration that is somewhat higher than the stoichiometric concentration required for aluminum combustion. For this concentration, the inter-particle distance is approximately 32 times the particle radius. The flame radius for aluminum particle is of the order of 2–5 times the particle diameter [17], hence there is little probability that group combustion will occur in case of aluminum particle combustion. Eckhoff [1] has reported on Leuschke's classical experiment which demonstrated the relative importance of radiative heat transfer in dust explosion flame propagation. In this experiment, on two sides of a double glass window, two dust clouds were simultaneously generated. Immediately after that one of the dust clouds was ignited using gas flame. It was observed from study that metallic (Zr, Ti, Al and Mg) flames produced sufficient radiation to ignite the cloud on the other side of the double walled glass window. Thus, it may be safely assumed during derivation of cloud scale energy and species source term that any single particle in the cloud will not "feel" the presence of other particles, however the flame will still be propagated primarily by radiation. This reasoning may be true for most of the metallic particles which have similar density as aluminum and lower volatility as compared to organic particles. Therefore the rate of consumption of particle phase (including radiation effects) could be transformed into quasi-homogeneous reaction rate for metallic particles according to following equation:

$$R''' = \alpha \times \frac{6}{d_p} \times R''(d_p, T, P, C_{O_2})$$

(2)

where α represents volume fraction of fuel, $R''(d_p, T, P, C_{O_2})$ represents the surface reaction rate, R''' represents the volumetric reaction rate, d_p is the particle diameter. Hence, in case of metallic dust explosion, surface reaction rate can be averaged over volume around particle and obtained volumetric reaction rate could be used in cloud scale CFD model. This will be the case, for instance, when the case of Al-air dust explosion would be considered.

For the case of metallic particles, then, the problem of connecting it to the dust cloud CFD model, as described in Fig. 1, would be to develop an appropriate mathematical description of the R term in Eq. (2) above, and then use it as a volumetric reaction rate term in a gas-phase CFD to describe the explosion in a "dust cloud" gas. Of course it will be assumed that the particles are in the micron range, and hence in Stokes' regime of settling with a small particle relaxation time, so that the particles actually follow the gas with fidelity.

As described earlier, the R term in Eq. (2) could be either estimated through a detailed analytical model incorporating the particle scale heat and mass transport, and reaction chemistry.

Indeed such an exercise has been accomplished by the authors but is being communicated through a different paper which will be dedicated to that model alone. Here, we use the alternate approach, which is to incorporate the particle scale dynamics though an appropriate correlation which models the particle scale combustion. Such a correlation, widely applied for aluminum dust particle combustion is the Beckstead combustion time correlation [18]. The authors of the current contribution has shown elsewhere the Beckstead's correlation[18] also agrees very well with the single particle combustion model developed by them.

Use of Beckstead's Correlation for Aluminum Combustion

According to Beckstead [18], a correlation could be developed for combustion time of single Aluminum particle in various oxidizing environments based on combustion data from more than ten sources with 400 data points. The correlated expression is given as:

$$\tau_c = \frac{a d_{p0}^n}{X_{eff} T^{0.2} P^{0.1}} \text{ where } X_{eff} = y_{O_2} + 0.6 y_{H_2O} + 0.22 y_{CO_2} \tag{3}$$

where $a = 0.0244$ for $n = 1.5$ and $a = 0.00735$ for $n = 1.8$, pressure is in atmospheres, temperature in K, diameter in μm, and time in ms. Estimation of volumetric reaction rate based on Beckstead's correlation[18] involves assumption of the volumetric reaction rate and surface reaction rate in following forms respectively:

$$R''' = A e^{-E/RT} C_{Al}^x C_{O_2}^y \tag{4}$$

$$R'' = \frac{1}{\pi d_{p0}^2} \frac{d(\frac{\pi}{6} d_{p0}^3 \rho_{Al}/M_{Al})}{dt} \tag{5}$$

For CFD modeling of Al-air dust explosion using Arrhenius kinetics, therefore four volumetric reaction rate coefficients (A, E, x and y) needs to be estimated.

$$Ae^{-E/RT}C_{Al}^x C_{O_2}^y = \frac{\alpha}{M_{Al}d_{p0}^3}\frac{d(\rho_{Al}d_{p0}^3)}{dt} \approx \frac{\alpha}{Md_{p0}^3}\frac{\rho_{Al}d_{p0}^3}{\tau_c} \quad (6)$$

Applying Beckstead's combustion time correlation [18]

$$Ae^{-E/RT}C_{Al}^x C_{O_2}^y = \left(\frac{\alpha \rho_{Al}}{M_{Al}}\right)\frac{X_{eff}T^{0.2}P^{0.1}}{ad_{p0}^n} \quad (7)$$

In the right hand side the term inside parentheses represents effective aluminum concentration i.e. $C_{Al} = \alpha\rho_{Al}/M_{Al}$. Mole fraction of oxidizer species (X_{eff}) in the ambient will be independent of temperature and pressure (for the closed system) and will be proportional to oxygen concentration ($X_{eff} \propto C_{O_2}$) in case of Al-air dust explosion. Pressure of the closed system will be proportional to temperature: $P \propto T$. Applying the above three conditions:

$$Ae^{-E/RT}C_{Al}^x C_{O_2}^y = C_{Al}C_{O_2}T^{0.3}\varphi(d_{p0}) \quad (8)$$

$$\Rightarrow -E/RT = 0.3 \ln T + \ln(\varphi(d_{p0})/A) \quad (9)$$

Assuming the reaction of aluminum-air is first order with respect to both the components (in absence of any better information) and the second term in RHS of Eq. (9) is independent of temperature, the activation energy could be evaluated from the slope of plot of $(0.3 \ln T)$ v/s. $1/T$. Hence,

$$E = -0.3R\frac{d(\ln T)}{d(1/T)} \quad (10)$$

Applying Eq. (10), the "effective" activation energy of 8.55 kJ/mol is obtained at 3400 K (approximate of Al-air flame temperature). The activation energy for Al particle surface reaction is reported as 73.6 kJ/mol [19]. The low value of activation energy seems to indicate that combustion is diffusion limited in this case, with a certain degree of falsification of the intrinsic kinetics. Consequently, the validity of Beckstead's correlation [18] is limited to large scale aluminum particles (particle size range of 15–760 μm) only from which it is derived.

Organic Particles

Eckhoff [1] in his book referred to the work of Gieras et al. [20], and reported that the maximum flame radius of coal particle is of the order of 3–6 times the radius of the particle. The result also shows that flame radius increase with increasing in volatile composition of the coal particle. Law [21] reported the maximum ratio of flame radius to radius of heptane droplet is approximately 15. Therefore, this can be argued that the ratio of flame radius to particle radius increases with increase in volatile content or the particle.

Organic particles mainly consist of volatiles including water vapor, fixed carbon and ash. Along with these components pores may be also present in the particles. Therefore organic particles are having bulk density which is lower than the true density of the fixed carbon and ash. The bulk density is defined as:

$$\rho_{bulk} = \varepsilon_{FC} \rho_{FC} + \varepsilon_{Ash} \rho_{Ash} + \varepsilon_{volatile} \rho_{volatile} + \varepsilon_{pore} \rho_{gas} \quad 11)$$

$$1 = \varepsilon_{FC} + \varepsilon_{Ash} + \varepsilon_{volatile} + \varepsilon_{pore} \Rightarrow \varepsilon_{FC} + \varepsilon_{Ash}$$

$$= (1 - \varepsilon_{pore}) - \varepsilon_{volatile} \quad 12)$$

Here ε_i and ρ_i represents volume fraction and true density of component 'i': fixed carbon, ash content, volatile, pores and gas. The true density of fixed carbon (high strength graphite) and the ash content (mineral such as alumina, silica and magnesia.) is reported as 1500 kg/m³ and 2600–3900 kg/m³ respectively [22]. The absolute density of most of the organic liquids/volatile lies between 800 and 1000 kg/m³ [23] much lower than the fixed carbon and ash content. Therefore assuming the volume fraction of pore (ε_{pore}) constant, if low density volatile is present in the organic particle it will reduce the volume of higher density fixed carbon and ash content and hence the presence of volatiles reduces the bulk density of the particle. If all the volatiles are removed by heating from the particle, the pore volume fraction of the particle will increase. As the density of gases is order of magnitude lower than the fixed carbon and ash content the bulk density of particle will reduce in this case. According to Fig. 3, the inter-particle distance reduces with decrease in density, therefore for very volatile particles, the individual particle combustion mode may not be observed. It is probable that in many cases burning of individual particle will be felt by neighboring particles. Therefore in case of organic particle combustion, estimation of quasi-homogeneous reaction rate similar to metallic dust cloud will give erroneous result and hence should not be applied. For organic particle combustion reaction rate will be determined from one set of cloud combustion and thereafter its predictability validated for other set of conditions.

CFD MODEL FOR DUST EXPLOSION AND PARAMETER ESTIMATION

CFD Model Formulation

The model formulation has been discussed and the methodology has been demonstrated using hydrogen-air mixture combustion by Bind et al. [14]. As mentioned by Bind et al. [14], the density-based solver of commercial CFD software Fluent 6.3.26 has been used for modeling dust explosion. For modeling chemistry during dust explosion species transport and finite-rate chemistry model has been used. A dust-air mixture approach, which is best suited for very small particles with a small particle relaxation time, is applied in present study. Accordingly kinetic parameters and physical properties are modified to account for combustible solid particles in the dust cloud. The density-based solver of commercial CFD software Fluent 6.3.26 has been used for modeling dust explosion which solves the governing equations of continuity, momentum, energy, and species transport simultaneously:

Continuity: $\dfrac{\partial \rho}{\partial t} + \nabla \cdot (\rho \vec{v}) = 0$ (13)

$$\dfrac{\partial}{\partial t}(\rho \vec{v}) + \nabla \cdot (\rho \vec{v}\vec{v}) = -\nabla p + \nabla \cdot (\bar{\bar{\tau}}_{eff}) + \rho \vec{g}$$ (14)

$$\dfrac{\partial}{\partial t}(\rho E) + \nabla \cdot (\rho \vec{v} E + p\vec{v}) = \nabla \cdot (k_{eff} \nabla T - \sum_j h_j \vec{J}_j + \bar{\bar{\tau}}_{eff} \vec{v}) + S_h$$ (15)

$$\dfrac{\partial}{\partial t}(\rho Y_i) + \nabla \cdot (\rho \vec{v} Y_i) = -\nabla \cdot ((\rho D_{i,eff} + \mu_t/Sc_t)\nabla Y_i) + R_i'''$$ (16)

The following equation for turbulent kinetic energy (k) and turbulence dissipation are solved after Eqs. (13), (14), (15) and (16) at each step

$$\dfrac{\partial}{\partial t}(\rho k) + \nabla \cdot (\rho \vec{v} k) = \nabla \cdot \left(\left(\mu + \dfrac{\mu_t}{\sigma_k}\right)\nabla k\right) + G_k + G_b - \rho\varepsilon - Y_M + S_k$$ (17)

$$\dfrac{\partial}{\partial t}(\rho\varepsilon) + \nabla \cdot (\rho \vec{v}\varepsilon) = \nabla \cdot \left(\left(\mu + \dfrac{\mu_t}{\sigma_\varepsilon}\right)\nabla\varepsilon\right) + C_{1\varepsilon}\dfrac{\varepsilon}{k}(G_k + C_{3\varepsilon}G_b)$$
$$- C_{2\varepsilon}\rho\dfrac{\varepsilon^2}{k} + S_\varepsilon$$ (18)

The equations mentioned above contain effective physical properties and kinetic parameters. These effective properties depend upon physical properties of each phase, components and dust loading in combustible dust cloud. Equation for estimation of some of the effective physical properties and kinetic parameters are mentioned in Table 1. Cloud density and pressure are related using a modified equation of state [24]. The reason for this modification is due to the presence of solid particles which does not exert any pressure on the enclosure containing dust cloud at steady state. Another main modification required is estimation of volumetric reaction rate from single particle combustion which is mentioned in previous section. In the proposed model stoichiometric combustion has been applied as it is intended to model the worst consequence of dust explosion for risk and hazard analysis application purpose. Although the stoichiometric combustion conditions may not be always achieved in actual practice, it does provide worse consequence prediction and hence a limiting envelope for different non-stoichiometric conditions leading to different levels of incomplete combustion.

Table 1: Effective physical properties and kinetic parameters for CFD Model

Reference	Physical properties and kinetic parameters	Relationship
Wallis [24]	Cloud density (ρ)	$\rho = \left(1 + \alpha[\frac{\rho_s}{\rho_g} - 1]\right) \frac{p}{RT\sum_i (Y_i/M_{w,i})}$ $Y_i \rightarrow$ mass fraction of vapor phase of species 'i'
Einstein's equation	Cloud viscosity (μ)	$\mu = \mu_g(1 + 2.5\alpha)$
Maxwell's equation	Cloud thermal conductivity (k)	$\frac{k}{k_g} = 1 + \frac{3\alpha}{(k_s + 2k_g)/(k_s - k_g) - \alpha}$
Wallis [24]	Cloud ratio of specific heat (γ)	$\gamma = (Cp_g + Y_s Cp_s)/(Cv_g + Y_s Cv_s)$
–	Cloud volumetric reaction rate (R''')	$R''' = Ae^{-E/RT} C_{Fuel}^m C_{O_2}^n = \alpha \times (6/d_p) \times R''$ (Heterogeneous/homogeneous single particle reaction rate $\rightarrow$ quasi homogeneous reaction rate)

The estimated effective physical properties and kinetic parameters are applied in CFD model as if the dust cloud is an "effective" gas phase. In the earlier work of Bind et al. [14], the methodology has been demonstrated using gaseous mixture combustion in 20 L Siwek apparatus. In present study the model has been extended to include dust combustion (explosion) in the similar apparatus, treating the gas–solid mixture as a quasi-homogeneous phase.

As pointed out in Section 2, depending on whether the dust explosion is for a metallic powder or organic powder, the way in which the particle-scale events are incorporated in the CFD model would be different. In fact, a somewhat different algorithm needs to be utilized in each case. Fig. 4 summarizes this approach, and in fact is an algorithmic representation of the overall modeling approach shown in Fig. 1. In the rest of this contribution, the central idea shown in Fig. 4 has been detailed in specific cases of aluminum and starch powder explosion.

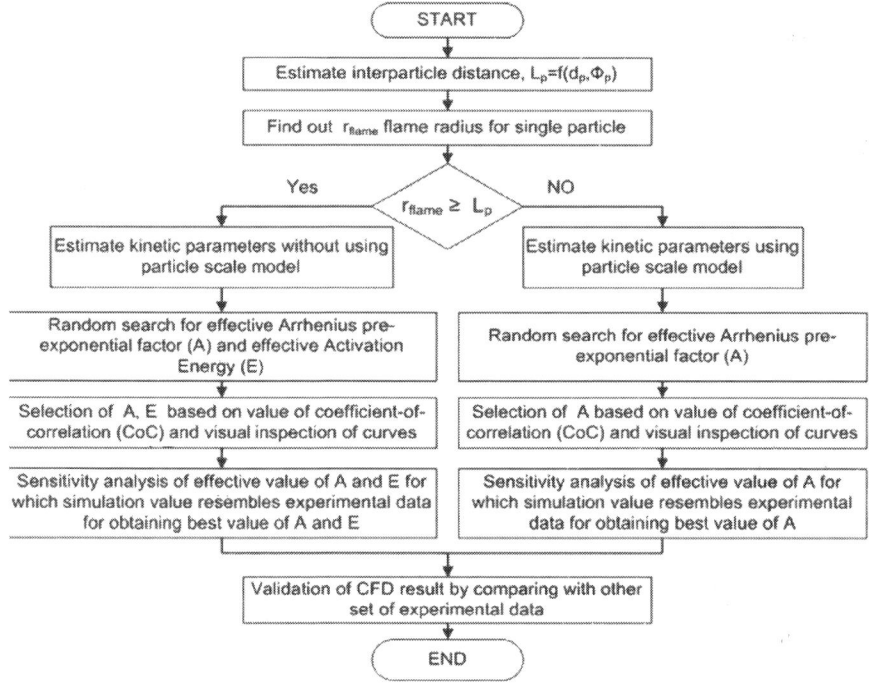

Figure 4: Solution algorithm for dust cloud kinetic parameter estimation.

Observed kinetic Parameter Estimation based on Experimental Data

The actual dust explosion is a complex phenomenon, spatial and temporal variations of most of the parameters (pressure, temperature, etc.) are nonlinear. Therefore there was a need of standard experimental apparatus whose data could be used to predict the real dust explosion. 20 L dust characterization apparatus emerged as standard apparatus worldwide due to lower capital and operating cost and consistency of result with larger apparatus [25]. In this standardized setup, the pressure history of the vessel is obtained as an output. Three dimensional symmetry of spherical geometry reduces the requirement of modeling it completely (owing to symmetry considerations), therefore only one-eighth of the 20 L sphere has been modeled in current studies. As representative samples for metallic and organic dust, aluminum and starch particles has been chosen, respectively.

Parameter Estimation for Aluminum Dust Explosion

Experimental data provided in Ogle's Ph.D. thesis [13] has been principally considered for estimation of kinetic parameter using CFD model and model validation purposes. The paper published by Dufaud et al.[5] has been also considered to verify the applicability of estimated parameters from particle scale model in different sets of experimental results of various researchers. The works mentioned above are in 20 L apparatus only. Few large scale dust explosions are also mentioned in literature; however enough data is not provided in them to carry out validation of CFD model. In the current CFD approach, the kinetic expression has been assumed similar to Ogle's model [13]. However in this case activation energy estimated from particle scale model has been used, as per the discussion around Beckstead's correlation in Section 2.1.1.

In the context of metallic dust particles, it is important to mention the parallel approach (to our work) that has been followed by Goroshin et al. [26], [27] and [28]. Goroshin et al. [27] has stressed that representation of polydisperse dust by some average particle size is not adequate due to variation of particle combustion rate with particle size. Discrete combustion has been suggested for aluminum dust explosion by Goroshin et al. [28] as it was observed that for lean dust mixtures, the flame width is comparable to or even less (in average) than the distance between particles. However, in our opinion modeling of all the underlying mechanisms (kinetics versus transport) at the same time and considering discrete modeling will make it a complex and, computationally costly. Besides, such detailed turbulence-chemistry interactions may not be needed for modeling dust explosions for most safety applications. This is particularly relevant in view of the uncertainties in the experimental data as there is no data available at present that correlates particle scale combustion with distribution in particle size, and the variation in controlling mechanism (kinetics or heat and mass transport) of combustion.

Thus, in the present study, the entire particle size range has been modeled assuming the mean particle size as the representative size of the particle. It is assumed that the particles with same mechanism of combustion have equal observed activation energy and could be directly incorporated from particle scale combustion model. However the observed Arrhenius pre-exponential factor, needs to be estimated from one set of experimental data (pressure versus time in a closed vessel) for that sample only, to account for the effect of polydisperse dust.

A particle size dependent "effective" Arrhenius pre-exponential factor (A) has been estimated for Al-air using experimental data available in literature. Variation of this parameter does not change the combustion mechanism of combustion, only time of combustion is altered. The value of A has been evaluated by random search of pre-exponential factor (A) in a large domain followed by selection of value of 'A' based on value of coefficient of correlation (CoC) and visual inspection of curves. The CFD modeling of Al-air dust explosion has been carried out using activation energy of $E = 8.55 \times 10^{03}$ kJ/kmol (approximately 316 kJ/kg) obtained from particle scale model in previous section. The mixture was ignited by patching temperature of 3500 K in sphere (radius 3.0 cm) at the center. The model setting and initial condition mentioned in Table 2 were used for simulation.

Table 2: Model settings and initial parameters for Al-air simulation

Model settings		Initial condition in vessel (corresponding to 500 g/m³ aluminum suspension in air)	
Space	3D		
Time	Unsteady, 1st-Order Implicit	Mass fraction of Al	0.2797
Viscous	$k\text{-}\varepsilon$ turbulence model	Mass fraction of Al_2O_3	0
Heat transfer	Enabled	Mass fraction of O_2	0.189
Species-transport	Reacting (Al, Al_2O_3, O_2, N_2)	Mass fraction of N_2	0.5313
Time step (s)	10^{-5} s	Temperature	300 K
Max. iterations per time step	100		

Reaction rate term for species:

Al: $-r_{Al} = A * \exp(-8.55e+06 \text{(J/kmol)}/RT) * C_{Al} * _cO_2$.

$O_2 :-_rO_2 = 0.75 \times A^*\exp(-8.55e+06\text{(J/kmol)}/RT) * C_{Al} * _cO_2$.

$Al_2O_3 : _rAl_2O_3 = 0.5 \times A * \exp(-8.55e+06\text{(J/kmol)}/RT) * C_{Al} * _cO_2$.

$N_2 :_rN_2 = 0$.

Parameter Estimation for Starch Dust Explosion

Starch is an organic material and hence undergoes Group Combustion, as explained in Section 2.2. The experimental data provided in Ph.D. thesis of Dahoe [29], for 15 μm particle diameter corn starch in a 20 L apparatus, has been used for estimation of kinetic parameters and validation of CFD model in present work. Large scale experimental work, in a 10.3 m³ cylindrical vessel, carried out by Kumar et al. [6] has been considered in next section.

For the cases where group combustion of particles occurs instead of single particle combustion, particle scale models are not available or not reliable due to complexity involved with combustion mechanism. Also in such cases, the material characterization or data used in model is not available. However, for purposes of modeling, the observed kinetic parameters could be estimated for particular dust-air system based on experimental data on the overall combustion characteristics. To extract the effective kinetics via numerical regression of the experimental data, two parameters (effective Arrhenius pre-exponential factor, A, and effective Activation Energy, E) has been selected for analysis.

The method of evaluation of observed kinetic parameters has been mentioned in Fig. 4. Note that in this case, the regression process itself involved a complete CFD calculation with Fluent 6.3.26 in the 20 L spherical geometry, to simulate the experiments of Dahoe [29]. Essentially, by this process one is claiming that whatever uncertainties exist between the model and real experiment, are all lumped into the two unknown parameters (A and E). The model, initial condition, and reaction kinetics were used for testing the feasibility of simulation is presented in Table 3. Subsequently, these parameters were used to simulate other cases of starch dust explosions, as discussed later.

Table 3: Model settings and initial parameters for starch-air simulation

Model settings		Initial condition in vessel (this corresponds to cornstarch concentration of 429 g/m³ or equivalence ratio 1.876)	
Space	3D	Mass fraction of starch	0.25
Time	Unsteady, 1st-order implicit	Mass fraction of H_2O	0
Viscous	k–ε turbulence model	Mass fraction of CO_2	0
Heat transfer	Enabled	Mass fraction of O_2	0.15
Species-transport	Reacting (5 species: starch, H_2O, CO_2, O_2, N_2)	Mass fraction of N_2	0.6
Time step (s)	5.0×10^{-5} s	Temperature	300 K

Max. Iterations per time step	100		

Reaction rate term for species:

starch: $-r_{Starch} = 1.8e^{+08} (m^3/(kmol\ s)) \times \exp(-100\ (kJ/mol)/RT) * C_{Starch} * _cO_2$.

O_2: $-r_{Starch} = 6 \times 1.8e^{+08}(m^3/(kmol\ s)) \times \exp(-100(kJ/mol)/RT) * C_{Starch} * _cCO_2$.

H_2O: $_rH_2O = 5 \times 1.8e^{+08}(m^3/(kmol\ s)) \times \exp(-100(kJ/mol)/RT) * C_{Starch} * _cO_2$.

CO_2: $_rCO_2 = 6 \times 1.8e^{+08}(m^3/(kmol\ s)) \times \exp(-100(kJ/mol)/RT) * C_{Starch} * _cO_2$.

N_2: $_rN_2 = 0$.

RESULTS AND DISCUSSION

Simulation of Aluminum Dust Explosion

For simulating the case of Aluminum dust explosion, incorporating the particle scale information in the dust-gas CFD simulation, the results were compared with Ogle et al.'s [30] experimental data and simulation work. Ogle [13] in his Ph.D. work developed a one-dimensional model based on spherical symmetry. Fig. 5 shows the experimental and theoretical work of Ogle [13], wherein the two show clearly very different trends. We believe this discrepancy emanates from the inadequacy in description of particle-scale effects in that original work [13]. Following kinetics was used in the two models:

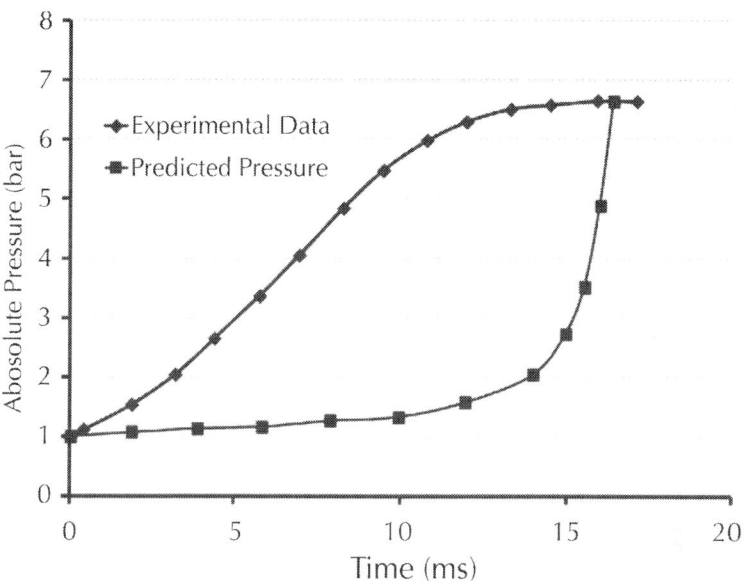

Figure 5: Experimental data and theoretical model comparison for Al (d_p = 12.1 μm and C_{Al} = 300 g/m³) combustion in 20 L apparatus, as reported by Ogle et al. ([30]).

Ogle's model [13]:

$$-r_{Al} = 1.4e+05((m^3/g)^{1/2}\ s^{-1}) * \exp(-24(kcal/mol)/RT) \\ * \rho^{3/2} Y_{Al} * Y_{O_2}^{1/2} \quad (19)$$

Current CFD simulation :

$$-r_{Al} = A * \exp(-8.55e+06(J/kmol)/RT) * C_{Al} * C_{O_2} \quad (20)$$

where A = 3200 m³/(kmol s) for particle size of 6.69 ± 1.1 μm
= 1600 m³/(kmol s) for particle size of 12.1 μm
= 800 m³/(kmol s) for particle size ($d_{P.50}$) of 11 μm

In our case, the "effective" Arrhenius factor "captures" the particle scale effects, through the treatment described in Section 2.1.1.

As shown in Fig. 6, CFD results are qualitatively and quantitatively similar to experimental results, and remarkably capture the experimentally observed trends of Ogle [13]. Fig. 6 also shows a comparison of CFD simulation result with recently published experimental data of Dufaud et al. [5]. Again, our CFD simulation results (incorporating the particle scale effects) closely tracks the experimental trends, in contrast to the prediction of Ogle et al. [30] where the trends could not be captured. It is noteworthy that in both the cases it was assumed that all the aluminum particle particles in the 20 L apparatus start burning at the same time. Ogle [13] considers heterogeneous surface reaction (activation energy ~100 kJ/mol) as the rate determining step whereas Dufaud et al. [5] considers both heterogeneous surface reaction (activation energy ~12 kJ/mol) as well as species diffusion limitation for single particle combustion. It is worth noting that in the present CFD simulations, the effective observed activation energy of $E = 8.55$ kJ/mol was considered whereas Dufaud et al. [5] considers activation energy ~12 kJ/mol for modeling heterogeneous surface reaction of aluminum particles. It is clear that these two values are of the same order of magnitude, indicating strong mass transfer limitations (and the same underlying physics) in both the cases. Naturally, our model is general enough to capture whatever controlling regime may exist in the combustion process.

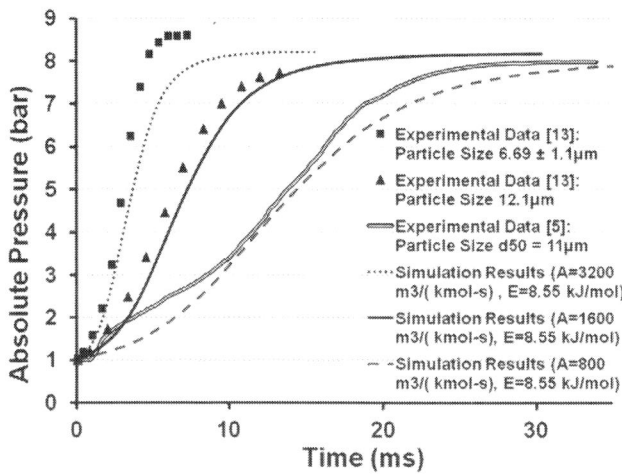

Figure 6: Comparison of CFD simulation with Ogle's [13] and Dufaud et al. [5] experimental work.

Simulation of Starch Dust Explosion

In this case, CFD simulations were performed without using an explicit particle scale description, as outlined in Sections 2.2 and 3.2.2. For these cases, determination of effective kinetics has been carried out using one set of experimental data, whereas validation has been done for different sets of experimental data of 20 L dust explosion apparatus available in literature. Value of coefficient of correlations (CoCs) and visual inspection of curves are the basis of validation study. For validation purposes, two different studies were undertaken: one for the trend analysis and another for the maximum pressure achieved. As shown inFig. 7, the CFD results are qualitatively and quantitatively similar to experimental results reported in the doctoral thesis of Dahoe [29]. However, the predicted final overpressure is slightly higher than the experimental data. This is expected since the actual experiments in 20 L standard Siwek apparatus involve unaccounted heat loss, incomplete combustion etc. which is not considered in simulation. The experimental results were provided in Dahoe's thesis [29] only for 15 μm starch particle size therefore only one value for effective Arrhenius pre-exponential factor, A, could be obtained. The final effective reaction rate expression used in the CFD simulations is:

$$-r_{Starch} = 1.8e^{+08}(m^3/(kmol\ s)) \times \exp(-100(kJ/mol)/RT) * C_{Starch} * C_{O_2} \quad (21)$$

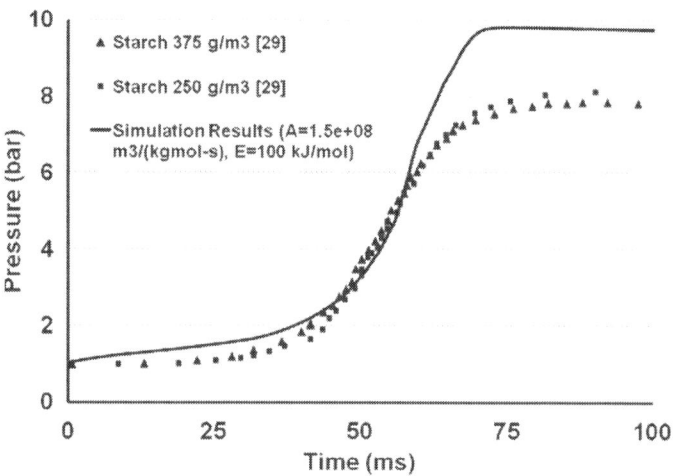

Figure 7: Comparison of simulation results with experimental data of Dahoe [29].

The estimated observed activation energy for starch-air explosion in 20 L apparatus is of the same order of magnitude as that for H_2-air estimated by Bind et al. [14]. The same order of magnitude for observed activation energy suggests homogeneous combustion for dust explosion phenomenon of starch-air mixture. The reason for homogeneous-gaseous type combustion of starch-air mixture might be due to requirement of much lower gasification temperature for starch compared to flame temperature of starch-air mixture during dust explosion phenomenon. Therefore, before the actual burning of starch, it is decomposed into smaller gaseous molecules, mixing of gaseous fuel and oxidizer takes place, and finally homogeneous combustion takes place; although some minor amount of char is also present during combustion process. In case of aluminum-air dust explosion presented in current study such homogeneous combustion is not observed. The reason for this is relatively larger particle combustion duration of micron size aluminum solid dust particles compared to gaseous fuel molecules formed during starch combustion. The composition of aluminum reduces slowly with reduction of mass of aluminum during particle burning process.

Case Study: Modeling Practical Process Plant Situation

This case study deals with validation of dust explosion model for organic particle dust explosion frequently occurring in agriculture sector. In this study, experimental work of Kumar et al. [6] involving corn starch dust explosion experiments in a 10.3 m³ cylindrical vessel, has been simulated using the developed model. The experiment was performed in a 1.5 m diameter and 5.7 m high (H/D = 3.8) cylindrical vessel whose schematic is shown in Fig. 8. For details of experimental setup and procedure Kumar et al. [6] work could be referred.

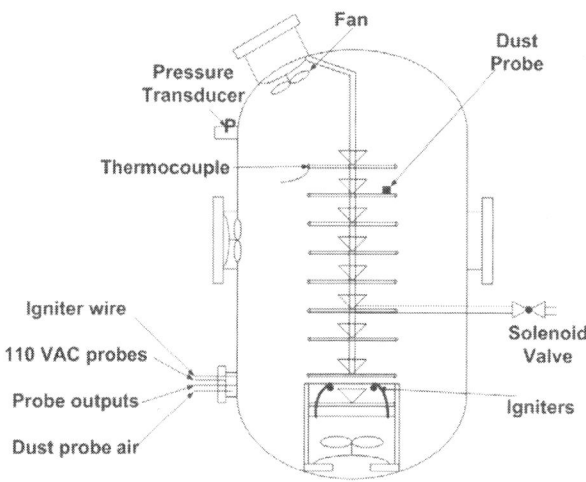

Figure 8: Schematic of the experimental apparatus (Kumar et al. [6]).

The experiments in the paper of Kumar et al. [6] were performed at ambient conditions: ~25 °C (298 K) and ~100 kPa at the time of ignition of dust cloud. The approximate average size of starch particles were 20 μm. Pressure histories was recorded for varying degrees of turbulence. Maximum pressure and rate of pressure rise for corn starch-air dust explosion was claimed to be consistent with literature in this study. The experimental result of the Kumar et al.'s [6] describing average pressure history in the vessel with and without fan generated turbulence for corn starch- air mixture is given in Fig. 9. As expected, in the vessel, the rate of pressure rise increases with increased turbulence generated due to fans compared to condition when fan is off.

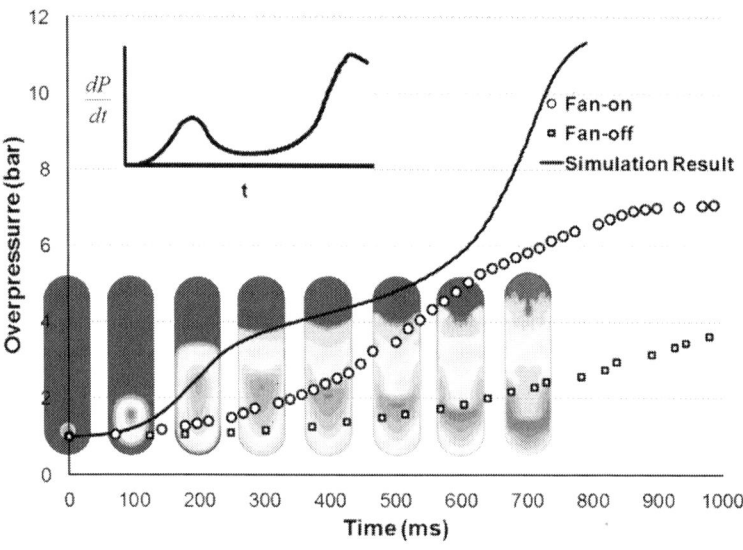

Figure 9: Average pressure history for cornstarch (291 g/m^3)/air mixture dust explosion in 10.3 m^3 experimental setup (Kumar et al. [6]) and simulation results.

The model, initial condition, and reaction kinetics were used for testing the feasibility of simulation is presented in Table 4. The mixture was ignited by patching temperature of 3500 K in sphere (radius 25.0 cm) at the bottom of the vessel and kept at mid-distance between axis and wall of the cylinder. Dynamic grid adaptation technique based on gradient of temperature has been used for simulation. Same kinetics, as reported in Section 4.2 (Eq. (21)), were used for these simulations. No other parameters were arbitrarily set. However, it may be noted that in Dahoe [29], the data is reported for 15 μm starch particles whereas in Kumar et al.'s [6] experimental data is for 20 μm starch particles. In the absence of available kinetic parameters for 20 μm starch particles, kinetic data for 15 μm starch particles evaluated in previous section were applied in present study without any further changes.

Table 4: Model settings and initial parameters for case study simulation

Model settings		Initial condition in vessel (This corresponds to cornstarch concentration of 0.429 kg/m³ or equivalence ratio 1.876)	
Space	3D		
Time	Unsteady, 1st-Order implicit	Mass fraction of starch	0.25
Viscous	$k-\varepsilon$ turbulence model	Mass fraction of H_2O	0
Heat transfer	Enabled	Mass fraction of CO_2	0
Species-transport	Reacting (5 species: starch, H_2O, CO_2, O_2, N_2)	Mass fraction of O_2	0.15
		Mass fraction of N_2	0.6
Time step (s)	1.0×10^{-5} s	Temperature	300 K
Max. iterations per time step	100		

Reaction rate term for species:

Starch: $-r_{Starch} = 1.8e^{+08}$ (m³/(kmol s)) × exp(−100(kJ/mol)/RT) * C_{Starch} * $_cO_2$

O_2: $-r_{Starch} = 6 \times 1.8e^{+08}$(m³/(kmol s)) × exp(−100(kJ/mol)/RT) * C_{Starch} * $_cO_2$.

H_2O: $_rH_2O = 5 \times 1.8e^{+08}$(m³/(kmol s)) × exp(−100(kJ/mol)/RT) * C_{Starch} * $_cO_2$.

CO_2: $_rCO_2 = 6 \times 1.8e^{+08}$(m³/(kmol s)) × exp(−100(kJ/mol)/RT) * C_{Starch} * $_cO_2$.

N_2: $_rN_2=0$.

One of the main observed differences between experimental data and simulation is total combustion time of starch-air mixture in cylinder. According to the simulation results shown in Fig. 9, the flame reaches upper wall of the cylinder in approximately 800 ms. This combustion time is lower than the experimentally predicted value of combustion time by Kumar et al. [6]. One of the main reasons for reduction in predicted combustion time is use of effective kinetic parameters of 15 μm starch particles instead of 20 μm used in experiments. The combustion kinetics for 15 μm starch particles in air is expected to be markedly faster than that of 20 μm starch particles. Fig. 9 shows the contours of temperature for starch-air dust explosion, obtained using CFD modeling in cylinder representing experimental setup. After ignition near the bottom of cylinder, the flame moves rapidly towards the unreacted starch-air mixture. The movement of flame is due to reaction as well as due to expansion of burned section caused by increased temperature and increase in the number of moles of gases.

Another major difference between experimental observation and simulation result is the trend of the average pressure history curve which could be observed in Fig. 9. Since dust explosion is deflagration phenomenon therefore the pressure in the vessel could be considered uniform for all practical purposes, hence average pressure is compared with the experimental data. Fig. 9 shows that according to experimental data, during initial stages of dust explosion, pressure and rate of pressure rise in the cylinder increases smoothly with time and in final stage of combustion, maxima is observed for pressure and rate of pressure rise starts decreasing and reaches to zero value when pressure becomes maximum. Simulation results also confirm a similar trend as shown, except for the reduction in the rate of pressure rise in between ignition and the instant of combustion completion in the cylinder. The reason for this discrepancy is evident from the contour plots in Fig. 9, which show that the incipient flame rapidly moves in both directions until it reaches the bottom. Therefore the initial rate of pressure rise in cylinder is due to combustion in two different planes. The combustion in lower direction stops at a time of about 200–300 ms due to consumption reactant exhaustion in the lower portion. The period 200–300 ms is also the time after ignition at which a decrease in rate of pressure rise is observed for the first time.

The third difference in predicted value and simulation results is maximum pressure reached in cylinder during dust explosion. In the present simulation, the peak pressure observed is of the order of 10.0 bar whereas Kumar et al. [6] reported a value of approximately 7.0 bar. The possible reason for difference might be use of different starch dust particle (15 μm instead of 20 μm used in experiment) combustion kinetics in simulation and use of adiabatic boundary condition at the vessel wall. Further, 15 μm starch particle might be more efficient for combustion compared to 20 μm starch particles which is used by Kumar et al. [6] for experimental study. It was reported by Kumar et al. [6] that incomplete combustion was observed during starch-air dust explosion. In an another experimental data of corn starch-air dust explosions, Radandt et al. [31] has reported maximum pressure of 10.47 bar for H/D ratio of 4 where corn starch median size was 15 μm. H/D ratio of 4 by Radandt et al. [31] is close to the geometry of experimental vessel used by Kumar et al. [6] for which H/D ratio is 3.8. Thus, our results are in reasonable agreement with that data as well.

The present case study suggests that pressure in the process vessel rises in finite time for starch-air dust explosions. Therefore, with the application of developed CFD tool various mitigation measures, such as suppression, venting and partial inerting, could be designed and optimized. Apart from designing one mitigation measure at a time, a combination of above mentioned techniques could be designed and optimized with the application of CFD.

SUMMARY AND CONCLUSIONS

The present study was concerned with modeling of dust explosions in a 20 L spherical apparatus (the so-called "Siwek 20 L apparatus), which is most widely used for dust explosion characterization as per international standards. The well-defined experimental procedure and sufficient data available in literature for 20 L spherical apparatus makes it appropriate for validation of developed CFD model for dust explosions. Due to complexity and computational costs involved with combined turbulent multiphase flow and combustion during dust explosion, a dust-gas mixture approach has been used to model the cloud scale phenomena in present study. This assumption have helped in obtaining better prediction of dust explosion compared to existing options at the same time keeping the scope for future improvements.

The density based solver of the commercial Finite Volume based CFD software, Fluent 6.3.26, was used for CFD modeling of combustion/dust explosion in 20 L apparatus. In this model, the "dust cloud" is assumed as homogeneous mixture which comprises of solid dust particles and gaseous oxidizers. Therefore, the CFD model mentioned above incorporates effective physical properties and kinetic parameters which depend upon physical properties of each phase, components and dust loading in combustible dust cloud. The equation for estimation of some of the effective physical properties and kinetic parameters are presented. Algorithm for estimation of observed kinetic parameters for dust explosion has been presented for two different types of dust particles explosions: metallic dust particle explosion, and organic dust particle explosion. The variation of approach for observed kinetic parameter estimation helps in avoiding the complexity and computational cost arises due to group combustion of organic particles and helps in establishing a practical approach for using CFD in dust explosion consequence prediction. The CFD modeling using particle scale model has been demonstrated for aluminum dust explosion, whereas for organic particle combustion CFD modeling of dust explosion is carried out for starch without using particle scale model. In both the cases, either using, or without using particle scale model for CFD simulation, model has been validated against experimental data. Qualitatively and quantitatively similar results were obtained compared to experimental results for both aluminum and starch combustion.

The estimated observed kinetic parameters for starch-air dust explosion has been applied for modeling of dust explosions in realistic conditions. Case study dedicated to organic dust explosion has been presented. The CFD simulation of starch-air dust explosion is validated with experimental data and it was demonstrated that the using same physics and estimated kinetic parameters using 20 L apparatus, fairly satisfactory prediction in real geometries could be obtained. Thus, the numerical challenge here is no longer to model the transport-chemistry, but ability to mesh large volumes and complex geometries. This has been successfully shown by CFD modeling of case study. Therefore, the developed CFD model could effectively utilized for design of safety systems and risk assessment of real dust explosions.

REFERENCES

1. R.K. Eckhoff, Dust Explosions in the Process Industries, 3rd ed., Gulf Professional Publishing, 2003.
2. P. Field, Dust Explosions, Elsevier, 1982.
3. K.N. Palmer, Dust Explosions and Fires, Halsted Press, New York, 1973.
4. P. Field, Explosibility assessment of industrial powders and dusts, Dept. of the Environment, Building Research Establishment, 1983.
5. O. Dufaud, M. Traoré, L. Perrin, S. Chazelet, D. Thomas, Experimental investigation and modelling of aluminum dusts explosions in the 20 L sphere, J. Loss Prev. Process Ind. 23 (2010) 226–236.
6. R.K. Kumar, E.M. Bowles, K.J. Mintz, Large-scale dust explosion experiments to determine the effects of scaling on explosion parameters, Combust. Flame 89 (1992) 320–332.
7. T. Skjold, Review of the DESC project, J. Loss Prev. Process Ind. 20 (2007) 291–302.

8. T. Skjold, B.J. Arntzen, O.R. Hansen, O.J. Taraldset, I.E. Storvik, R.K. Eckhoff, Simulating dust explosions with the first version of DESC, Process Saf. Environ. Prot. 83 (2005) 151–160.
9. T. Skjold, B.J. Arntzen, O.R. Hansen, I.E. Storvik, R.K. Eckhoff, Simulation of dust explosions in complex geometries with experimental input from standardized tests, J. Loss Prev. Process Ind. 19 (2006) 210–217.
10. A.E. Dahoe, R.S. Cant, B. Scarlett, On the decay of turbulence in the 20-liter explosion sphere, Flow Turbulence Combust 67 (2001) 159–184.
11. A.E. Dahoe, K. Hanjalic, B. Scarlett, Determination of the laminar burning velocity and the Markstein length of powder-air flames, Powder Technol. 122 (2002) 222–238.
12. A.E. Dahoe, Nat K. van der, M. Braithwaite, B. Scarlett, On the effect of turbulence on the maximum explosion pressure of a dust deflagration. KONA – powder and particle 19 (2001) 178–196.
13. R.A. Ogle, A new strategy for dust explosion research: a synthesis of combustion theory, experimental design and particle characterization, The University of Iowa, 1986.
14. V.K. Bind, S. Roy, C. Rajagopal, CFD modelling of dust explosions: rapid combustion in a 20 L apparatus, Can. J. Chem. Eng. 89 (2011) 663–670.
15. K. Annamalai, W. Ryan, S. Dhanapalan, Interactive processes in gasification and combustion – Part III: Coal/char particle arrays, streams and clouds, Prog. Energy Combust. Sci. 20 (1994) 487–618.
16. C. Crowe, M. Sommerfeld, Y. Tsuji, Multiphase Flows with Droplets and Particles, CRC Press, Boca Raton, 1998.
17. P.E. DesJardin, J.D. Felske, M.D. Carrara, Mechanistic model for aluminum article ignition and combustion in air, J. Propul. Power 21 (2005) 478–485.
18. M.W. Beckstead, Correlating aluminum burning times, Combust. Explos. Shock Waves 41 (2005) 533–546.

19. J. Bouillard, A. Vignes, O. Dufaud, L. Perrin, D. Thomas, Ignition and explosion risks of nanopowders, J. Hazard. Mater. 181 (2010) 873–880.
20. M. Gieras, R. Klemens, S. Wojcicki, Ignition and combustion of coal particles at zero gravity, Acta Astronaut. 12 (1985) 573–579.
21. C.K. Law, Combustion Physics, Cambridge University Press, New York, 2006.
22. M.F. Ashby, D.R.H. Jones, Engineering Materials. 1. An Introduction to their Properties and Applications, 2nd ed., Butterworth-Heineman, Oxford, 2002.
23. R.K. Sinnott, Chemical Engineering, 3rd ed., Chemical Engineering Design, Butterworth-Heineman, Oxford, 1999. vol. 6.
24. G.B. Wallis, One-Dimensional Two Phase Flow, McGraw Hill, New York, 1969.
25. R. Siwek, Determination of technical safety indices and factors influencing hazard evaluation of dusts, J. Loss Prev. Process Ind. 9 (1996) 21–31.
26. S. Goroshin, J. Mamen, A. Higgins, T. Bazyn, N. Glumac, H. Krier, Emission spectroscopy of flame fronts in aluminum suspensions, Proc. Combust. Inst. 31 (2) (2007) 2011–2019.
27. S. Goroshin, M. Kolbe, J.H.S. Lee, Flame speed in a binary suspension of solid fuel particles, in: Proceedings of the Combustion Institute, vol. 28, 2000, pp. 2811–2817.
28. S. Goroshin, J.H.S. Lee, Y. Shoshin, Effect of the discrete nature of heat sources on flame propagation in particulate suspensions, in: Twenty-Seventh Symposium (International) on Combustion/the Combustion Institute, 1998, pp. 743–749
29. A.E. Dahoe, Dust Explosion: A Study of Flame Propagation, TU Delft, 2000.
30. R.A. Ogle, J.K. Beddow, L.D. Chen, P.B. Butler, An investigation of aluminum dust explosions, Combust. Sci. Technol. 61 (1988) 75–99.

31. S. Radandt, J. Shi, A. Vogl, X.F. Deng, S.J. Zhong, Cornstarch explosion experiments and modeling in vessels ranged by height/diameter ratios, J. Loss Prev. Process Ind. 14 (2001) 495–502.

Chapter 7

Correlations for Flame Speed and Explosion Overpressure of Dust Clouds inside Industrial Enclosures

M. Silvestrini[a], B. Genova[a], and F.J. Leon Trujillo[b]

[a]Direzione Centrale Emergenza, Corpo Nazionale dei Vigili del Fuoco, Via Cavour 5, 00184 Roma, Italy
[b]Facoltà di Ingegneria, Università degli Studi di Roma "La Sapienza", Via A. Scarpa 16, 00161 Roma, Italy

ABSTRACT

Explosion relief vents on enclosures in powder-handling plants are currently designed according to technical standards that in some situations may overestimate the required vent area significantly.

These technical standards sometimes do not take into account the real work conditions of industrial plants (e.g. turbulence intensity) and therefore explosion worst cases are not always foreseeable. The availability of methods either for the evaluation of explosion overpressure or sizing of relief vents, with involvement of the pre-ignition turbulence, could be very useful for a better estimate of these quantities. In this work two empirical correlations are presented: the first one allows the calculation of the flame speed and the burning velocity starting from the explosion indices K_{St} and P_{max} of the standardized 20-l sphere test. The second allows either the calculation of the explosion overpressure or the sizing of relief vents of an enclosure.

INTRODUCTION

In the first section of this paper the concepts of laminar and turbulent burning velocity and flame speed are introduced together with a basic description of current measurement techniques. A presentation follows of the empirical correlations developed to predict the burning velocity and flame speed of a given dust–air mixture taking into account the pre-ignition turbulence.

Laminar Burning Velocity in Dust Clouds

The laminar burning velocity, S_{CL}, is the speed at which a plane flame front will propagate into a stationary, quiescent flammable mixture of infinite extent (Drysdale, 1998). For gaseous mixtures, S_{CL} is considered a fundamental property of a given fuel–air mixture at a specified pressure and temperature. Its value is exactly defined into the mixture continuum and depends on fuel nature and mixture concentration. Several ways are available for the measurement of laminar burning velocity. Andrews and Bradley (1972) provide an excellent review of the field. Methods for determination of the burning velocity fall into one of two groups: methods based on propagating flames, such as propagating flames in cylindrical tubes and freely propagating spherical flames in closed bombs; methods based on stationary flames, such as Bunsen burner flames and nozzle burner flames. The "spherical bomb" method (Lewis & von Elbe, 1987) allows the evaluation of the laminar burning velocity by the pressure–time record and flame speed measurement in the course of the isochoric combustion. Another widely used method is that of "cylindrical tube". The tube is filled with the flammable mixture and after ignition a flame front starts to travel upward. The speed and shape of the travelling flame are then estimated from video recordings. The mass conservation principle applied to the mixture flowing through the flame front leads finally to the calculation of the laminar burning velocity. Tubes "bottom open"–"top close" and vice versa are both used. When the tube is closed at the bottom (bottom ignition), the expansion of burnt gases must be taken into account since it largely contributes to the speed of the flame front. In the field of stationary flames the Bunsen burner technique allows to evaluate the laminar burning velocity starting from the measurement of the flame cone angle and velocity of the issuing flow. Mention of flat flame burners equipped with porous disc (Botha & Spalding, 1954) should be done. In this case the laminar burning velocity equals the velocity of the issuing flow. Laminar burning velocity and flame thickness are intrinsic properties of gaseous mixtures, but the same point of view raises some doubts if applied to dust–air mixtures. Fig. 1 illustrates that the average distance between particles of size 15 µm and particle density 1000 kg/m^3, in a dust cloud with concentration 500 g/m^3, is about 150 µm (Dahoe, Hanjalic, & Scarlett, 2002). The cloud so formed appears as a homogeneous mixture at macroscale level but indeed it is a discrete system made of a three-dimensional assembly of particles immersed in a continuum of air. Since the average interparticle distance of a dust–air mixture is some order of magnitude larger than the intermolecular distance for gases, chemical and physical properties of discrete dust–air system cannot be regarded as average properties of the mixture itself. Consequently the laminar burning velocity originating from those chemical and physical properties could not be assumed as a fundamental property of the mixture. Despite this, Bradley and Lee (1984) underlined the concept that a cloud of dust exhibits laminar burning velocity and flame thickness similar to those of gaseous mixtures if the dust particles are able to produce an appreciable quantity of volatile matter when flowing into the pre-heat-zone of the flame. This volatile matter mixes with the surrounding air producing a flammable mixture ready to burn into the flame-reaction zone.

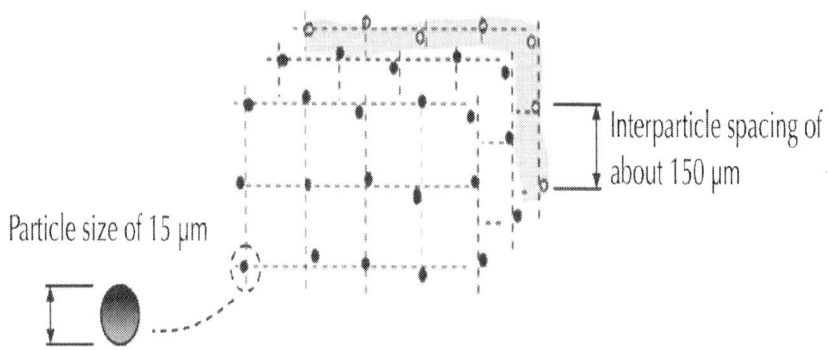

Figure 1: Dust cloud made of a finite number of particles immersed in a continuum of air.

The measurement of the laminar burning velocity of dust–air mixtures is performed with the same techniques used for gases but the results are strongly variable depending on method and apparatus (see Table 1 and Table 2).

Table 1: Laminar burning velocity of Lycopodium

Lycopodium	Method	Concentration (g/m³)	S_{cl} (m/s)
Proust and Veyssiere (1988)	Square duct 200×200 mm²	35–100	0.47
Glinka et al. (1996)	Cylindrical tube d=160 mm	35–200	0.69
Gieras et al. (1995)	Square duct 80×80 mm²	30–230	0.47
van der Wel (1993)	Burner	150–450	0.30

van der Wel (1993)	20-l sphere	1000	0.17
Kaesche-Krischer (1959)	Burner	180–500	0.25
Pedersen and van Wingerden (1995)	Cylindrical tube d=128 mm	50–175	0.41
Krause et al. (1996)	Cylindrical tube d=60 mm	160–710	0.28
Krause et al. (1996)	Cylindrical tube d=100 mm	180–635	0.50

Table 2: Laminar burning velocity of maize starch

Maize starch	Method	Concentration (g/m³)	S_{CL} (m/s)
Proust (1993)	Square duct 200×200 mm²	235	0.27
van der Wel (1993)	Burner	400–800	0.20
van der Wel (1993)	20-l sphere	400	0.13
Mazurkiewicz and Jarosinski (1991)	Burner	–	0.13
Mazurkiewicz and Jarosinski (1994)	Square duct 50×50 mm²	260–760	0.14
Pedersen and van Wingerden (1995)	Cylindrical tube d=128 mm	75–200	0.59
Krause et al. (1996)	Cylindrical tube d=60 mm	370–1200	0.22
Krause et al. (1996)	Cylindrical tube d=100 mm	80–430	0.40
Dahoe et al. (2002)	Tube+burner (flat flame)	330	0.29

Source: Krause and Kasch (2000).

About Lycopodium dust the use of spores from various species could also contribute to the large spread in the results summarized in Table 1 (Skjold, 2003).

However, the most of the spread in the results for both Lycopodium and maize starch is to ascribe to the nature of dusts themselves. In fact size and mass of dust particles are some order of magnitude higher than those of gas molecules. As consequence, dust particles exhibit high settling velocity and therefore specific test conditions are needed to keep them suspended in the air. If the test is performed in a constant volume bomb, a strong airflow is necessary to disperse the dust before ignition. The cloud so formed is highly turbulent and therefore the burning velocity measured is highly turbulent too. Increasing of ignition delay time, time between beginning of dispersion and triggering of ignition, results in lower turbulence intensity of the cloud and, hence, lower turbulent burning velocity. The laminar burning velocity is then evaluated by extrapolating at "zero turbulence" (Bradley, Chen, & Swithenbank, 1988; Tai, Kauffman, & Sichel, 1988). Regarding the method of cylindrical tube (Goroshin, Mamen, Lee, & Sacksteder, 2005; Han et al., 2000; Krause & Kasch, 2000; Pedersen & van Wingerden, 1995; Proust & Veyssiere, 1988) the technique of tube "bottom open"–"top close" with ignition at the bottom is widely used. The dispersion is obtained either by fluidized bed at the bottom of the tube or by vibrating sieve placed on the upper end. Frequently a small level of residual turbulence can be present due to the previous ascending airflow from the fluidized bed or to the wake of settling particles (Pedersen & van Wingerden, 1995). Sometimes acoustic standing waves take place in the tube leading to distortion of the flame front and growth in the rate of combustion (Eckhoff, 1997). Burner techniques (Dahoe et al., 2002; Horton, Goodson, & Smoot, 1977; Kolbe, 2001; Mazurkiewicz & Jarosinski, 1990) also highlight some problems. This method shows a clear dependence of the burning velocity on the burner diameter (Cassel, 1964; Kolbe, 2001). Moreover, it is necessary to resort to specific test set-up in order to keep the flame anchored at the burner rim. Mention of the sensitivity of dust flames to stretch and curvature affecting the burning velocity, especially with small diameter apparatuses, must be done (Dahoe et al., 2002). Last, but not least, the unsolved problem of generating clouds of uniform dust concentration inside test volumes. The picture depicted gives a clear idea about the difficulties encountered in performing a reliable determination of the laminar burning velocity of a dust–air mixture. The given summary and mainly the data provided in Table 1 and Table 2 show how the determination of the laminar burning velocity for a dust–air flame is not so univocal. In fact the current methods used are, at least, normally affected by the following factors: residual turbulence of the dispersion, wake turbulence of the settling particles (Pedersen & van Wingerden, 1995), instabilities of the flame front due to acoustic standing waves (Eckhoff, 1997), sensibility of the flame front to curvature (Dahoe et al., 2002), flame speed increased by buoyancy of burnt gases (Dahoe et al., 2002), non-uniform dust concentration inside test apparatuses (Wang, Pu, Jia, Gutkowski, & Jarosinski, 2006 reported difference of concentration up to 2 or 3 times along the tube height).

Turbulent Burning Velocity

Under laminar flow condition the flame front can be regarded as a thin reaction sheet separating the burnt gases from the unburnt mixture, but when the propagation of the flame front occurs under turbulent flow condition it becomes wrinkled, bent and distorted by turbulent eddies, so increasing its surface area. Locally, the combustion behaviour is still governed by the laminar burning velocity S_{CL} but the whole process proceeds with a turbulent velocity S_{CT} higher than the laminar one. The turbulent burning velocity depends on turbulence intensity u'_{rms} ("rms" stands for "root mean square") as shown (e.g.) by the following relationship (Veynante & Vervisch, 2002):

$$\frac{S_{CT}}{S_{CL}} = 1 + K\left(\frac{u'_{rms}}{S_{CL}}\right)^n, \quad (1)$$

where K and n depend, as a minimum, on fuel nature and mixture concentration.

Eq. (1) is only one of several empirical correlations for the turbulent burning velocity. The most of the gas–air mixtures have (Peters, 1992): 1 K 4 and 0.5 n 1.

The turbulence intensity quantifies the level of the random component of the velocity and it is defined as follows:

$$u'_{rms} = \sqrt{\overline{u'^2}} = \sqrt{\frac{\sum_1^N (u_i - \overline{u})^2}{N}}, \qquad (2)$$

where N is the total number of measurements, u_i is the velocity of "i" particle, $\overline{u}$ is the mean value of the velocity and u ' is the value of the instantaneous fluctuation. For dust–air mixtures several researchers have found a linear dependence of turbulent burning velocity on turbulence intensity u'_{rms} on the form:

$$S_{CT} = S_{CL} + Ku'_{rms}. \qquad (3)$$

Tezok, Kauffman, Sichel, and Nichols (1986) evaluated the velocity of combustion of two different maize starch–air mixtures, of concentration 300 and 700 g/m³, using a 0.95 m³ constant volume bomb. They found a linear relationship between the burning velocity and the rms turbulence intensity up to 7 m/s (seeFig. 2).

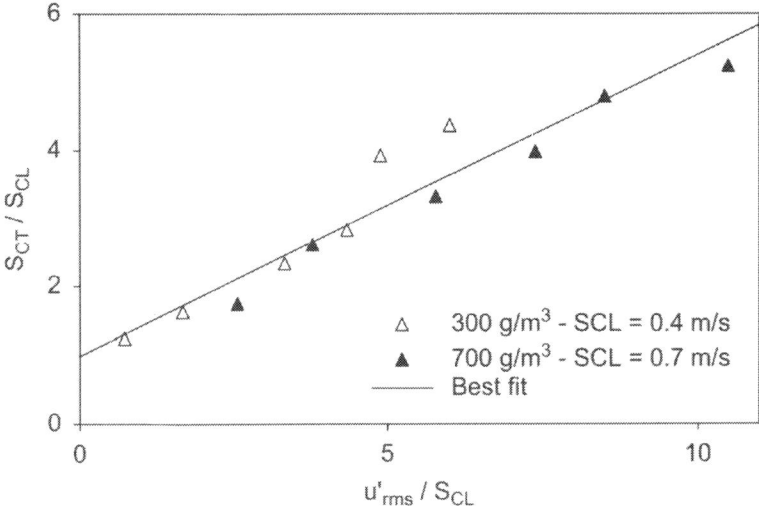

Figure 2: Normalized turbulent burning velocity of two maize starch–air mixtures (Tezok et al., 1986).

Pu, Jarosinski, Tai, Kauffman, and Sichel (1988), Pu, Lee, Kauffman, and Bernal (1989) and Pu, Jarosinski, Johnson, and Kauffman (1990) by means of tests performed with maize starch–air and aluminium–air mixtures in vessels of various capacity (from 6 to 950 l) showed the existence of a linear relationship between the turbulent burning velocity and the rms initial turbulence intensity (pre-ignition turbulence) up to 3.3 m/s. Van der Wel (1993) by explosion tests with the standardized 20-l sphere (equipped with the perforated dispersion ring) found a clear linear dependence of the turbulent burning velocity on the rms initial turbulence intensity up to 3 m/s. The dusts used were Lycopodium, activated carbon and potato starch. Gieras, Glinka, Klemens, and Wolanski (1995) and Gieras, Klemens, and Wolanski (1996) performed combustion tests with Lycopodium, wheat dust and maize starch in a 1.25 m³ spherical vessel. They found a linear relationship between the flame front speed and the rms pre-ignition turbulence intensity up to 10 m/s. Skjold (2003), thanks to measurements executed in a 20-l cubical vessel and a 20-l type USBM vessel (with rebound nozzle), obtained turbulent burning velocity (evaluated at the inflection point of the pressure–time history of combustion) linearly increasing with the rms turbulence intensity (at the same inflection point). He used niacin amid, Lycopodium, RDX and silicon dust. Pu, Jia, Wang, and Skjold (2007) performed constant volume combustion tests on both dust–air and gas–air mixtures in vessels of different shapes and volumes. They used the concept of maximum effective burning velocity, that is the burning velocity of an idealized flame with the same pressure–time trace of a real dust flame, and found:

- A near linear correlation, between the burning velocity and the initial turbulence intensity up to 3 m/s, almost independent on shape and volume of test vessel;
- Linear correlations, vessel dependent, between k_{st} values and the turbulence intensity at the time of ignition.

The latter suggests the possibility of a link between burning velocity and K_{St} value.

Concerning the influence of temperature and pressure on laminar burning velocity it can be expressed by an equation, generally adopted for both gases and dusts (Dahoe & de Goey, 2003; Gieras, Glinka, Klemens, & Wolanski (1995) and Gieras, Klemens, & Wolanski (1996); Metghalchi & Keck, 1982; van Wingerden, 1996), of the following kind:

$$S_{CL} = S_{CL}^0 \left(\frac{T_U}{T_0}\right)^\alpha \left(\frac{p}{p_0}\right)^\beta,$$

(4)

where S^0_{CL} is the laminar burning velocity at reference temperature and pressure (T_0 and p_0), α and β are empirical functions of the equivalence ratio and, finally, T_U and p are the temperature and pressure of the unburnt mixture (index "U" stands for unburnt, while the pressure p is the same for burnt gases and unburnt mixture). Reference values for the two exponents of Eq. (4) are: $\alpha=2.0$ and $\beta=-0.2$ (Hattwig & Steen, 2004). For dust–air mixtures $\alpha=2.0–1.7$ is normally assumed (Eckhoff, 1997; van Wingerden, 1996) and β=−0.5 to −0.36 (Dahoe, Zevenbergen, Lemkowitz, & Scarlett, 1996; Eckhoff, 1997). Considering a constant volume combustion, the expansion of burnt gases results in an adiabatic compression of unburnt mixture with temperature increase given by substituting Eq. (5) into Eq. (4) the following expression is obtained:

$$\left(\frac{T_U}{T_0}\right) = \left(\frac{p}{p_0}\right)^{(\gamma-1)/\gamma}. \tag{5}$$

$$S_{CL} = S^0_{CL}\left(\frac{p}{p_0}\right)^{\alpha-(\alpha/\gamma)+\beta}. \tag{6}$$

The choice of α=1.7, β=−0.36 and γ=1.4 leads to α−α/γ+β=0.126 that is very close to the value of 0.14 (van der Wel, 1993) assumed in this work for dust–air mixtures. So, the laminar burning velocity dependence on temperature and pressure can be described by the following equation:

$$S_{CL} = S^0_{CL}\left(\frac{p}{p_0}\right)^{0.14}. \tag{7}$$

The effects of temperature and pressure on the behaviour of the turbulent burning velocity have been not yet clearly defined and therefore any correction is made to the coefficient K of Eq. (3) in the present work. The turbulent burning velocity relationship here adopted can be obtained by substituting Eq. (7) into Eq. (3) that yields:

$$S_{CT} = S^0_{CL}\left(\frac{p}{p_0}\right)^{0.14} + Ku'_{rms}. \tag{8}$$

Formulation of the Problem

When a flammable mixture is ignited in the standardized 20-l sphere then a pressure–time history can be recorded as shown in Fig. 3. About the combustion processes developing in the test volume the following assumptions are made:

- The flame front is regarded as a thin reaction zone (Dahoe et al., 1996), although its thickness can reach several centimeters for dust–air mixtures;
- Turbulent length scales are disregarded also if their influence on the rate of combustion has been widely recognized. The reasons of the choice lie on the simplified nature of the present approach and minor importance of the length of turbulent structures with respect to turbulence intensity.

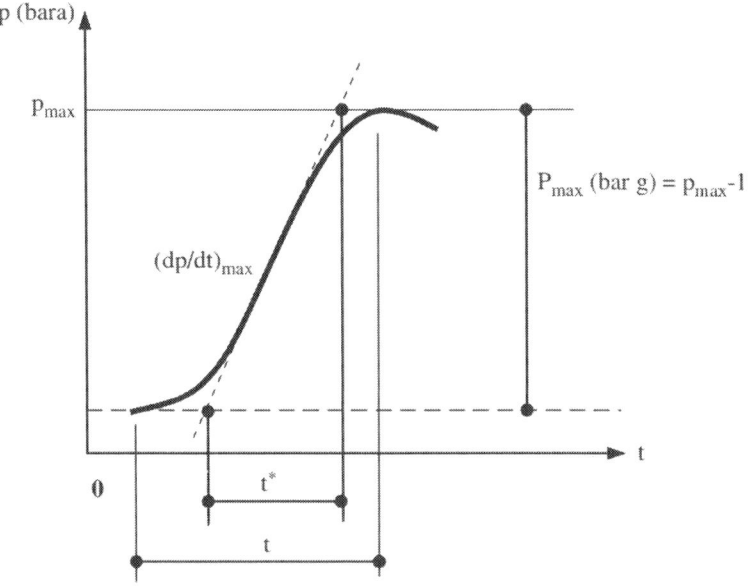

Figure 3: Qualitative pressure–time history of a flammable mixture in the standardized 20-l sphere.

Under these conditions the mean flame front speed can be written as $S_{Fmean}=r/t$, where r is the vessel radius and t the time needed to reach the vessel wall. The distance r covered by the flame front is proportional to the cube root of the vessel volume, $r \propto \sim V^{1/3}$, while the travelling time is proportional to the ratio of maximum overpressure, reached at the end of combustion, to maximum rate of pressure rise (see Fig. 3):

$$t \propto t^* = \frac{P_{max}}{(dp/dt)_{max}}.$$

By substitution of r and t, the previous expression of the flame speed becomes that arranged with the well-known "cube-root-law" (Bartknecht, 1981):

$$S_{Fmean} \propto V^{1/3} \frac{(dp/dt)_{max}}{P_{max}}$$

$$\left(\frac{dp}{dt}\right)_{max} V^{1/3} = K_{St} \qquad (9)$$

leads to the following proportionality relationship between flame speed and explosion indices K_{St} and P_{max} of a given dust–air mixture:

$$S_{Fmean} \propto \frac{K_{St}}{P_{max}}. \qquad (10)$$

After this step a suitable expression of the flame speed must be found. It can be supplied by the conservation of mass flow across a flame surface of area A:

$\rho_u S_C A = \rho_b S_F A,$

where S_C is the burning velocity and the indices "u" and "b" stand for "unburnt" and "burnt". The previous equation lead to the following expression of the flame speed:

$S_F = S_C \rho_u / \rho_b,$

where the ratio between densities of unburnt and burnt gases, E, is called expansion factor and it is a measure of the increase in gas volume produced by the combustion.

The previous equation can be shortened to the form (Harris, 1983) and it provides a link between the flame speed S_F and the burning velocity S_C via the expansion factor E of the burnt mixture. This relationship is valid under the following main conditions (Harris, 1983):

SF=SCE
- Planar, hemispherical or spherical flame;
- Laminar flow;
- Burnt gases always trapped behind the expanding flame front.

It also applies to pipes or ducts taking into account the increase in volumetric rate of combustion caused by increases in flame area.

Substitution of Eq. (8) into the last expression of the flame speed yields:

$$S_F = \left(S_{CL}^0 \left(\frac{p}{p_0}\right)^{0.14} + Ku'_{rms} \right) E. \tag{11}$$

Since, as underlined by Eq. (10), a proportionality relationship exists between the flame speed and the ratio K_{St}/P_{max}, the above Eq. (11) can be written as:

$$S_F = \left(S_{CL}^0 \left(\frac{p}{p_0}\right)^{0.14} + Ku'_{rms} \right) E \propto \frac{K_{St}}{P_{max}}. \tag{12}$$

The meaning of Eq. (12) is that two distinct terms contribute to the value of the rate of combustion expressed by the ratio K_{St}/P_{max}: the first depending on the laminar behaviour of the flame S_{CL}^0 and the second on its reactivity to turbulence K.

If the laminar term of the flame speed (Eq. (12)) is set equal to a fraction C_{Lam} of the ratio K_{St}/P_{max} then the following equation is obtained:

$$\left(S_{CL}^0 \left(\frac{p}{p_0}\right)^{0.14} \right) E = C_{Lam} \frac{K_{St}}{P_{max}}. \tag{13}$$

A similar treatment of the turbulent term of the flame speed (Eq. (12)) yields:

$$Ku'_{rms} E = \tilde{C}_{Turb} \frac{K_{St}}{P_{max}}$$

and assuming $C_{Turb} = C_{Turb}/u'_{mis}$ the following equation is obtained:

$$KE = C_{Turb}\frac{K_{St}}{P_{max}}.$$
(14)

The coefficients C_{Lam} and C_{Turb} can be at last calculated at the point of maximum overpressure of the pressure–time trace by Eqs. (13) and (14) recast on the following form:

$$C_{Turb} = KE_{min}\frac{P_{max}}{K_{St}}.$$
(15)

$$C_{Turb} = KE_{min}\frac{P_{max}}{K_{St}}.$$
(16)

Evaluation of the Coefficients of the Empirical Correlation for Turbulent Combustion of Dust–Air Mixtures

Knowledge of an adequate number of S^0_{CL} and K values of different dust–air mixtures and related explosion indices K_{St} and P_{max} (with reference to combustion tests performed in the standardized 20-l sphere equipped with perforated dispersion ring) allows the determination of coefficients C_{Lam} and C_{Turb} by Eqs. (15) and (16). The expansion factor, used in Eqs. (15) and (16), has been indicated as E_{min} because at the last stage of the explosion the adiabatic compression of unburnt and burnt gases yields (see Section 1.4):

$$E_{min} = \frac{\rho_u}{\rho_o} = \left(\frac{P_{max}}{p_0}\right)^{1/\gamma}.$$
(17)

The evaluation of the coefficients C_{Lam} and C_{Turb} by the way described above, using data of Table 3 and Table 4, yields (average values): $C_{Lam} = 0.01$ and $CTurb = 0.15$.

Table 3: Laminar burning velocity of selected dusts

N	Dust	Conc. (g/m³)	S_{CL} (m/s)	K_{St} (bar m/s)	P_{max} (bar g)	Reference/notes (primary sources of K_{St} and P_{max} data are: NFPA 68 (1994) and Eckhoff, 1997)
1	Lycopodium	60–175	0.41 max	129	6.7	Pedersen and van Wingerden (1995)/cylindrical tube d=128 mm (noise due to vibrating feeding system with rms turbulence intensity±0.4 m/s)
2	Lycopodium	200	0.26	155	8.5	Krause et al. (1996)/bunsen burner
3	Sulphur 20 μm	300	0.25	113	7	van der Wel (1993)/tube shaped burner φ=33 mm
4	Maize starch 15 μm	800	0.18	158	9.7	van der Wel (1993)/flat flame burner
5	Lycopodium	35–80	0.36	155	8.5	Proust (1993)/square duct 100 mm×100 mm
6	Maize starch	75–275	0.30 max	202/215	10.3/9.4	Proust and Veyssiere (1988)/square duct 200 mm×200 mm
7	Maize starch	330	0.29	202/215	10.3/9.4	Dahoe et al. (2002)/cylindrical tube with upper burner (flat flame)
8	Maize starch	330	0.288	202/215	10.3/9.4	Dahoe et al. (2002)/cylindrical tube with upper burner (flat flame)
9	Aluminium 10 μm	310	0.42	400	12.5	Ballal (1983)
10	Carbon (40% volatile) 12 μm	–	0.25	123	9.1	Ballal (1983)
						Vertical tube zero (g)
11	Carbon (Pittsburgh) 9 μm	–	0.33	100	6.73	Horton et al. (1977)
12	Poli Vinil Alcool	<200	0.25	128	8.9	Kaesche–Krischer (1959)
13	Maize starch	370	0.38	202/215	10.3/9.4	Krause et al. (1996)
						Cylindrical tube d=100 mm

14	Maize starch	235	0.27	202/215	10.3/9.4	Proust and Veyssiere (1988)/square duct 200 mm×200 mm
15	Lycopodium	200	0.42	129	6.7	Krause et al. (1996)/cylindrical tube d=100 mm
16	Aluminium	≈310	0.49	515/475	11.2/12	Goroshin et al. (2005)/tube apparatus ground-based testing—cylindrical tube—length 700 mm—φ=50 mm—the measured flame speed has been roughly divided by a factor 2 to calculate the burning velocity and then a correction has been applied to take into account the flame curvature
17	Titanium	≈415	0.53	265	8.3	Goroshin et al. (2005)/tube apparatus ground-based testing—cylindrical tube—length 700 mm—φ=50 mm—the measured flame speed has been roughly divided by a factor 2 to calculate the burning velocity and then a correction has been applied to take into account the flame curvature
18	Iron	≈725	0.18	65	6.5	Goroshin et al. (2005)/tube apparatus ground-based testing—cylindrical tube—length 700 mm—φ=50 mm—the measured flame speed has been roughly divided by a factor 2 to calculate the burning velocity and then a correction has been applied to take into account the flame curvature

19	Sulphur	–	0.29	113	7	Proust (2006)/tube apparatus—square duct—length 1500 mm–100 mm×100 mm²
20	Maize starch 14 μm	From 100 to 400 along tube height	0.29	158	9.7	Wang et al. (2006)/square duct apparatus—780 mm–160 mm×160 mm
21	Potato starch 26 μm	1000	0.17	60	7	van der Wel (1993)/flat flame burner
22	Lycopodium 33 μm	450	0.30	165	7.3	van der Wel (1993)/flat flame burner
23	Aluminium 10 μm	260	0.39	400	12.5	Cassel (1964)/bunsen burner φ=19 mm
24	Maize starch 15 μm	600	0.20	158	9.7	van der Wel (1993)/tube shaped burner φ=33 mm

Table 4: Values of K parameter of selected dusts

N	Dust	Concentration (g/m³)	K	Source
1	Maize starch	300/700	0.44	Tezok et al. (1986)
2	Lycopodium	70	0.37	Gieras et al. (1995/96)
3	Wheat dust	200	0.16	Gieras et al. (1995/96)
4	Maize starch	220	0.46	Gieras et al. (1995/96)
5	Coal	625	0.27	Zevenbergen (2004)
6	Potato starch	500/625	0.26	Zevenbergen (2004)
7	Potato starch	1000	0.32	van der Wel (1993)
8	Activated carbon	500	0.63	van der Wel (1993)
9	Lycopodium	500	0.72	van der Wel (1993)
10	Maize starch	400	0.55	van der Wel (1993)
11	Aluminium	250	0.75	van der Wel (1993)

The few available values of laminar burning velocity provided in Table 3 and used for determination of C_{Lam}, have been selected on the basis of the following criteria:

- The highest measured values of S_{CL}^0 obtained by tube apparatuses have been discarded because too high (probably due to residual turbulence and acoustic excitation);
- Some available results from spherical test vessel have also been discarded because of apparently too small values;
- Results of different test methods have been taken into consideration with the clear purpose of compensating possible drift of measured values related to a specific test method;
- Measures from flat flame burners have been considered, whenever available, because not affected by curvature, stretch, turbulence and buoyancy;
- Only measures from tubes and burners with appreciable cross section have been used in order to avoid correction of strongly stretched flame.

Fig. 4 and Fig. 5 show the values of C_{Lam} and C_{Turb} calculated using the described procedure together with their average values.

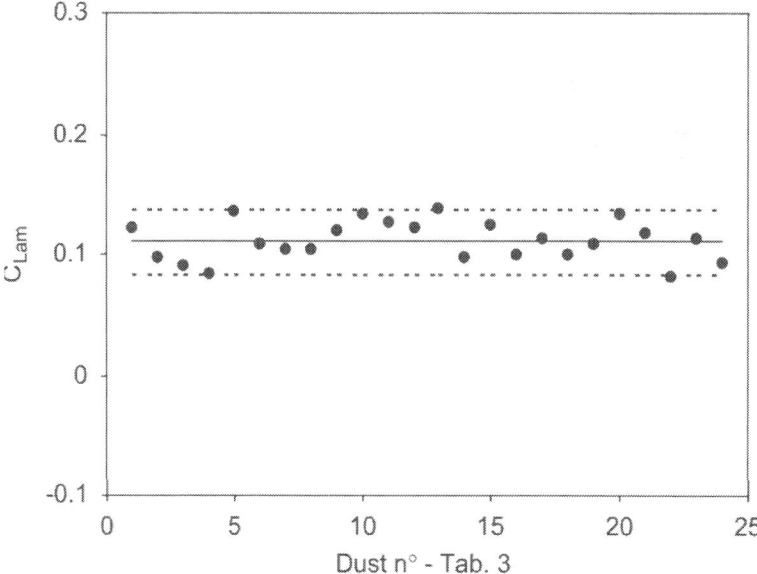

Figure 4: C_{Lam} calculated by Eq. (15).

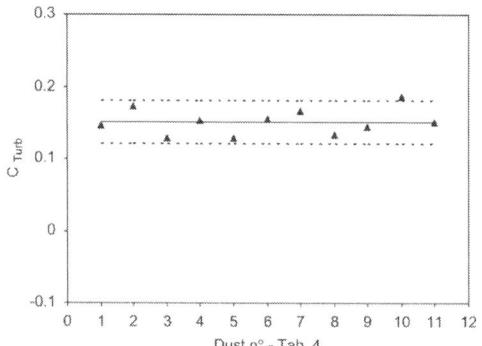

Fig 5: C_{Turb} calculated by Eq. (16).

Using the average values $C_{Lam} = 0:11$ and $C_{Turb} = 0:15$. It is possible, by Eqs. (13) and (14), to write the final relationships for the laminar burning velocity S^0_{CL} and parameter K (reactivity to turbulence) of a dust of known explosion indices K_{st} and P_{max} from standardized 20-l sphere:

$$S^0_{CL} = 0.11 \frac{K_{St}}{P_{max}(p_{max}/p_0)^{0.14} E_{min}},$$

(18)

$$K = 0.15 \frac{K_{St}}{P_{max} E_{min}}.$$

(19)

Fig. 6 and Fig. 7 show a comparison between measured–calculated values of laminar burning velocity and measured–calculated values of parameter K obtained by Eqs. (18) and (19). The good agreement between measured and calculated values of laminar burning velocity S^0_{CL} and reactivity to turbulence K is due to the use of the best fit coefficients $C_{Lam} = 0:11$ and $C_{Turb} = 0:15$ of Eqs. (18) and (19) obtained by the experimental data.

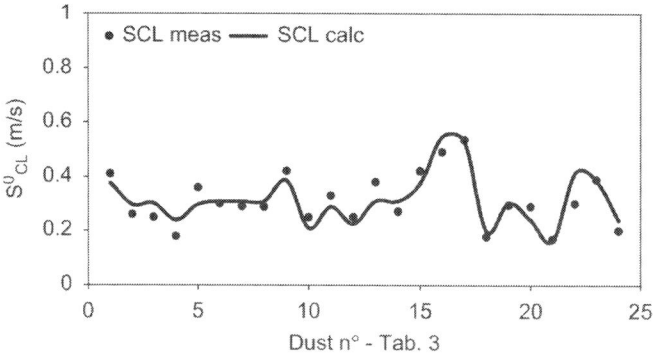

Figure 6: Laminar burning velocity—comparison between measured and calculated.

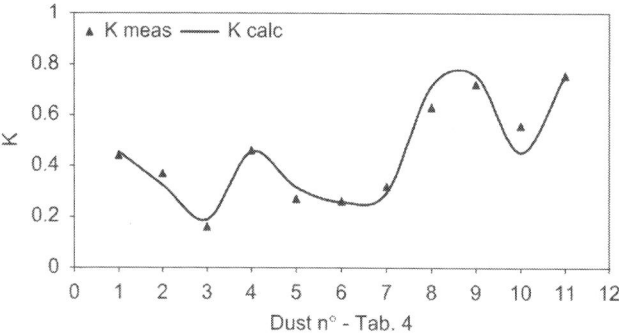

Figure 7: K parameter—comparison between measured and calculated.

Experimental data used for the comparisons are those of Table 3 and Table 4.

Application to gaseous mixtures of a procedure similar to that outlined above for dust–air mixtures leads to the following expression of laminar burning velocity:

$$S^0_{CL} = 0.41 \frac{K_G}{P_{max}(p_{max}/p_0)^{\alpha-\alpha/\gamma+\beta} E_{min}}, \tag{20}$$

where the explosion index K_{St} (for dusts) has been replaced by K_G (for gases).

The influence of temperature and pressure on laminar burning velocity of gas–air mixtures has been taken into account by means of (except for hydrogen and acetylene) α=2.0, β=−0.2 (Hattwig & Steen, 2004) andγ=1.4. This choice allows the calculation of the exponent of Eq. (6) that yields: α−α/γ+β=0.37.

For hydrogen and acetylene the following values of α and β have been used:

- Hydrogen α=1.26 (Milton & Keck, 1984), β=0.10 (Lewis, 1954);
- Acetylene α=2.0 (Milton & Keck, 1984), β=0.06 (Lewis, 1954).

The substantially different value of β for hydrogen and acetylene is due to changes of β versus laminar burning velocity. In fact β exhibits negative values for $S_{CL}^0 < 0.45 m/s$ and positive values for $S_{CL}^0 > 1.0 m/s$. (Lewis, 1954).

Fig. 8 shows a comparison between S_{CL}^0 measured and calculated using Eq. (20).

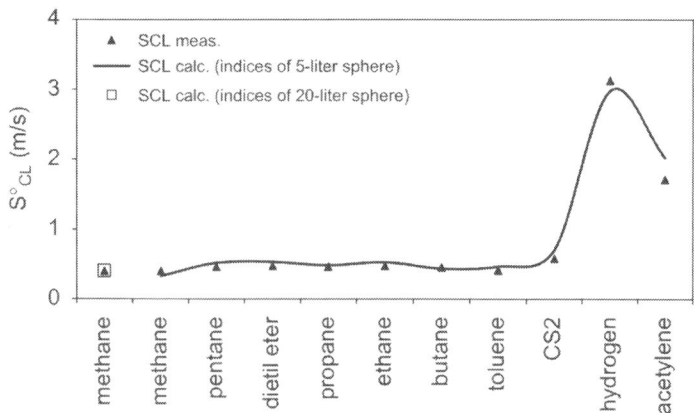

Figure 8: Laminar burning velocity of gaseous mixtures—comparison between measured and calculated values.

Empirical Correlations for Flame Speed, Laminar and Turbulent Burning Velocity of a Dust–Air Mixture

The behaviour of the expansion factor in the course of an explosion in the standardized 20-l sphere will be roughly discussed in this section, followed by a brief summary of the simplified relations obtained above for flame speed and burning velocity. As already stated the expansion factor expresses the increase in volume produced by combustion and is given by the ratio of densities of unburnt and burnt gases. Two distinct aspects of combustion contributes to the value of the expansion factor:

- The temperature increasing due to the exothermic reaction;
- The change in number of moles from reagents to products.

Neglecting the second contribution and taking into consideration an isobaric combustion at ambient pressure ($\rho_{uT0} = \rho_{bTad}$), the corresponding expansion factor E^0 can be expressed by the ratio of the flame temperature T_{ad} to the unburnt mixture temperature, T_0:

$$E^0 = \frac{\rho_u}{\rho_b} = \frac{T_{ad}}{T_0}. \tag{21}$$

A rough estimate of the expansion factor, for combustion at ambient pressure, shows that it is very close to the ratio between the maximum overpressure reached in an isochoric combustion and the ambient pressure (Bjerktvedt, Bakke, & van Wingerden, 1997; Proust, Roux, & Chhuon, 2000):

$$E^0 \cong \frac{P_{max} - P_0}{P_0} = \frac{P_{max}}{P_0}. \tag{22}$$

The numerical value of the above ratio is equal to the value of P_{max} (bar g) that is the maximum overpressure reached in the 20-l sphere test. On the other hand the expansion factor at the end of a constant volume combustion can be quickly estimated thinking that the density of burnt gases (at combustion end) equals the density of the unburnt mixture at the beginning of combustion $\rho_{b,f} = \rho_0$ and hence:

$$E_{min} = \frac{\rho_{u,f}}{\rho_{b,f}} = \frac{\rho_{u,f}}{\rho_0} = \left(\frac{p_{max}}{p_0}\right)^{1/\gamma}, \tag{23}$$

where $\rho_{u,f}$ and $\rho_{b,f}$ are the densities of unburnt and burnt mixture at the end of the isochoric combustion. In the course of the isochoric combustion the expansion factor changes gradually from its maximum value E^0 given by Eq. (22) to its minimum value E_{min} given by Eq. (23). A general expression of the expansion factor can be derived as follows:

$$E = \frac{\rho_u}{\rho_b} = \frac{T_b}{T_u} = \frac{T_{ad} + \Delta T}{T_0 + \Delta T}, \tag{24}$$

where ΔT is the temperature growth of the mixture caused by adiabatic compression. Adding and subtracting T_0 to numerator of Eq. (24) and dividing both numerator and denominator for T_0 the following relation is obtained:

$$E = \frac{T_{ad}/T_0 + ((T_0 + \Delta T)/T_0) - 1}{(T_0 + \Delta T)/T_0} = \frac{E^0 - 1 + (T_u/T_0)}{(T_u/T_0)}. \tag{25}$$

Since the equation of adiabatic transformation can be written as its substitution into Eq. (25), after some arrangements, leads to the final expression of the expansion factor:

$$\frac{T_u}{T_0} = \left(\frac{p}{p_0}\right)^{(1-1/\gamma)}$$

$$E = (E^0 - 1)\left(\frac{p}{p_0}\right)^{-(1-1/\gamma)} + 1. \tag{26}$$

Fig. 9 shows the change in expansion factor versus pressure calculated with Eq. (26).

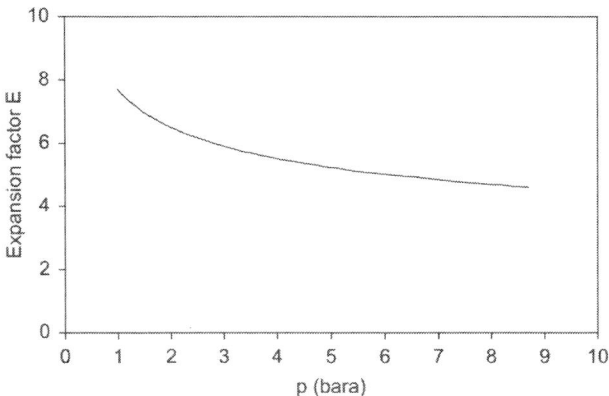

Figure 9: Change in expansion factor in the course of an isochoric combustion.

The flame speed of a dust–air mixture, taking into account pressure variation and pre-ignition turbulence, can be expressed as already seen (see Eq. (11)) by.

$$S_F = \left(S_{CL}^0 \left(\frac{p}{p_0}\right)^{0.14} + K u'_{rms} \right) E.$$

By substituting Eq. (18) (simplified expression of the laminar burning velocity S_{CL}^0) and (19) (parameter K) into the above relationship and restarting, the following simplified equation of flame speed is obtained:

$$S_F = \frac{K_{St}}{P_{max}(p_{max}/p_0)^{1/\gamma}} \left[0.11 \left(\frac{p}{p_{max}}\right)^{0.14} + 0.15 u'_{rms} \right] E. \tag{27}$$

where the expansion factor, pressure dependent, is calculated by Eq. (26). The burning velocity is then obtained from the previous equation by eliminating E:

$$S_C = \frac{K_{St}}{P_{max}(p_{max}/p_0)^{1/\gamma}} \left[0.11 \left(\frac{p}{p_{max}}\right)^{0.14} + 0.15 u'_{rms} \right]. \tag{28}$$

Fig. 10 and Fig. 11 show flame speed and burning velocity, calculated by Eqs. (27) and (28), versus pressure variation for a dust–air mixture with $K_{St} = 140\,\text{bar m/s}$ and $P_{max} = 9\,\text{bar g}$ (standardized 20-l sphere). Two levels of turbulence have been used in the calculation: $u'_{rms} = 0.5\,\text{m/s}$ and $u'_{rms} = 2.7\,\text{m/s}$.

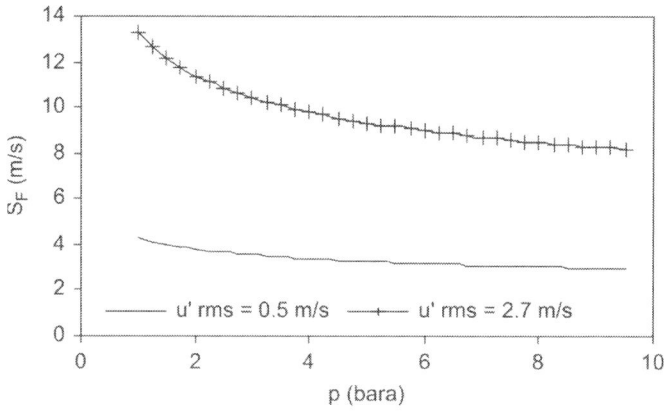

Figure 10: Flame speed of a dust–air mixture at two levels of pre-ignition turbulence. $-K_{St} = 140\,\text{bar m/s} - P_{max} = 9\,\text{bar g} - E^0 = 9$.

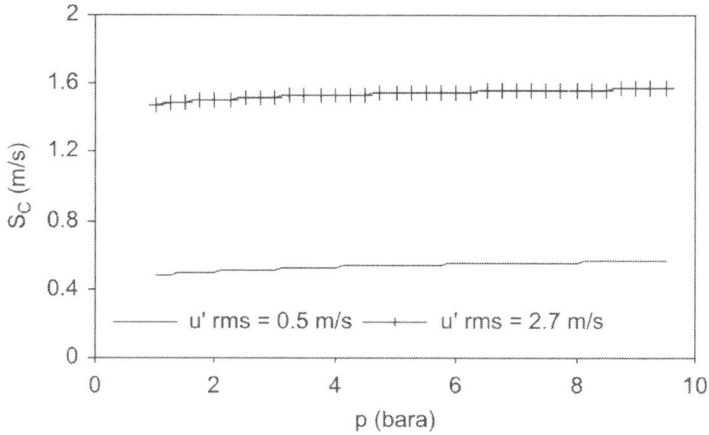

Figure 11: Burning velocity of the same mixture of Fig. 10.

Figure. 12 and Figure. 13 show a comparison between measured and calculated flame speed.

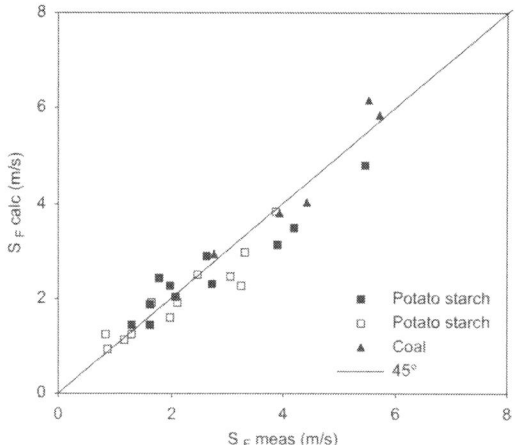

Figure 12: Flame speed—comparison between values measured (Zevenbergen, 2004) and calculated by Eq. (27).

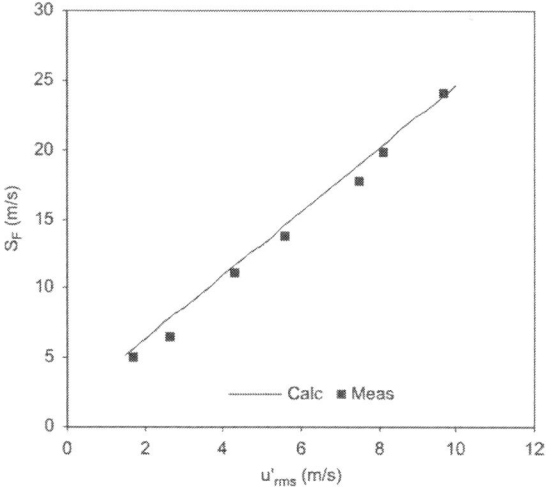

Fig 13: Flame speed comparison between values measured (maize starch 220 g/m^3—Gieras, Glinka, Klemens, & Wolanski (1995) and Gieras, Klemens, & Wolanski (1996)) and calculated by Eq. (27).

Figure. 14 shows a comparison between measured and calculated burning velocity.

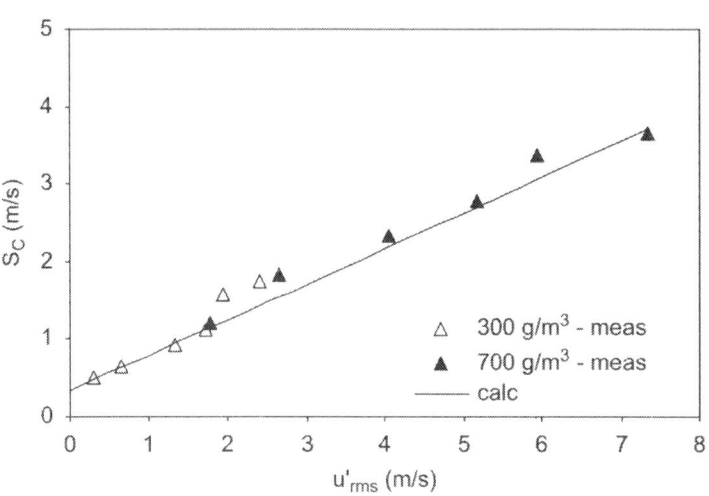

Figure 14: Burning velocity—comparison between values measured (Tezok et al., 1986—maize starch) and calculated by Eq. (28).

With reference to Fig. 12, the calculation of the flame speed has been done using Eq. (27) with the rms turbulence intensity supplied by Eq. (29) proposed by Dahoe, Cant, and Scarlett (2001) for the standardized 20-l sphere equipped with the rebound nozzle:

$$\frac{u'_{rms}}{u'^{0}_{rms}} = \left(\frac{t}{t_0}\right)^n, \qquad (29)$$

where $u'^{0}_{rms} = 3.75 \text{m/s}$, $t_0 = 0.06$ and $n = -1.61$ (for the perforated dispersion ring: $u'^{0}_{rms} = 2.68 \text{m/s}$, $t_0 = 0.06$ and $n = -0.06 \text{s}$ and $n = -1.49$).

Furthermore, with reference to Fig. 12 and Fig. 13, the flame speed has been calculated by Eq. (27) using the minimum value of the expansion factor $E_{min} = (p_{max}/p_0)^{1/\gamma}$ for the reasons explained below:

- First, in experimental works of Gieras, Glinka, Klemens, & Wolanski (1995) and Gieras, Klemens, & Wolanski (1996)

and Zevenbergen (2004), the sensors for detection of flame front were placed after half the radius of the respective test spheres. This means that, at the moment of detection, the internal pressure was greater than initial pressure and hence the expansion factor was in the range betweene^0 and E_{min};
- Second, the flame thickness can be very thick for dust–air mixtures. The consequence could be a combustion not yet completed at the moment of flame front arrival to the vessel wall (presence of pockets of unburnt material into burnt gases especially for high turbulence level). In this condition the expansion factor is low and probably very close to its minimum value E_{min} due to not immediate combustion of the fuel–air mixture.

Empirical correlations for flame speed and burning velocity based on the explosion indices of dust–air mixtures have been presented. These explosion indices are related to the standardized 20-l sphere (with perforated dispersion ring). The use of explosion indices from 20-l sphere equipped with the rebound nozzle requires some clarifications. Eckhoff reported that due to some problems with many cohesive dusts in the 20-l Siwek sphere, this "led to the development of an open nozzle system named a rebound nozzle which gradually replaced the original perforated ring. According to Siwek (1988), the new nozzle produces both maximum pressure and K_{St} values in reasonable agreement with those generated by the original perforated ring-system" (Eckhoff, 1997, pp. 531–532). The values of K_{St} to be used in Eqs. (18),(19) and (27) refer to the standardized 20-l sphere test with perforated ring system and therefore, if K_{St} values referred to the rebound nozzle are available then it is necessary to scale them in order to obtain the same S_{CL}^0 and K values calculated by Eqs. (18) and (19).

A way to determine the scaling factor could be by the evaluation of the ratio K_{St}/K_{St_reb} with the help of Eq (8) reported here:

$$S_{CT} = S_{CL}^0 \left(\frac{p}{p_0}\right)^{0.14} + Ku'_{rms}.$$

Since the flame speed is proportional to K_{St} (see Eq. (12)) and being the expansion factor and the maximum pressure the same for the two different nozzles, this leads to:

$$\frac{K_{St}}{K_{St\_reb}} = \frac{S^0_{CL}(p/p_0)^{0.14} + K2.68}{S^0_{CL}(p/p_0)^{0.14} + K3.75}.$$

Using a typical value of laminar burning velocity $S^0_{SL} = 0.25 \text{m/s}, K = 0.25$ and a maximum pressure of 9 bar, the previous equation yields: that could be assumed as an average value of the scaling factor:

$$\frac{K_{St}}{K_{St\_reb}} \simeq 0.80$$

This means that should be used either the 80% of K_{St_reb} (in place of K_{St}) or, if K_{St_reb} is entirely used, the coefficients C_{Lam} and C_{Turb} of Eqs. (18), (19) and (27) should be accounted for the 80% of their value. However, taking into account that, as reported by Siwek, the K_{St} values obtained with the two different nozzles are in reasonable agreement between them and the ratio K_{St}/K_{St_reb}, as estimated above, is not so far from unity, corrections could be avoided.

EXPLOSION OF DUST–AIR MIXTURES INSIDE FULL-SCALE INDUSTRIAL ENCLOSURES

Protection from explosions of dust–air mixtures is a complex matter. At present several vent sizing methods are available either as part of national or international guidelines or technical standards (e.g.NFPA 68, 1994; VDI 3673, 2002).

A comparative review of these methods together with experimental data about realistic dust explosions (performed in the decade 1980–1990 inside full-scale industrial enclosures), has been compiled by Eckhoff (1997). This review shows a heterogeneous and fragmentary situation related to volume, shape, characteristic turbulence level of the reviewed enclosures and their respective behaviour during the explosion. A further aspect to underline is that the size of vents evaluated with the methods mentioned above, particularly with VDI 3673 (2002) and NFPA 68 (1994), is normally overestimated (Eckhoff, 1997). A conservative attempt to treat the matter on a more rational basis, with regards to the pre-ignition turbulence, has been done with the 2007 edition of NFPA 68 (Zalosh, 2007). The new issue of NFPA 68takes explicitly into account the initial turbulence intensity of the enclosure to be protected. In fact the 2007 edition of NFPA 68 gives a vent area multiplier of 1.7 (that envelopes the 90% of data) for turbulence intensity less than 2 m/s and introduces an additive vent area, that linearly depends on turbulence, for turbulence intensity higher than 2 m/s. Perhaps this allows the calculation of the right vent size for high turbulence intensity but still increases the overestimation of those cases already overestimated with the previous issue of the NFPA 68 (1994).

Eckhoff (1997) classifies the main components of industrial plants according to volume, length to diameter ratio (L/D), internal congestion and function:

- Large empty enclosures of L/D<4 ;
- Large slender enclosures (silos) of $L/D>4$;
- Smaller, slender enclosures of $L/D>4$;
- Intermediate (10–25 m^3) enclosures of small L/D;
- Cyclones;
- Bag filters;
- Elongated enclosures of very large L/D (tubes, pipelines, and galleries).

In the present work an empirical correlation for the evaluation of the explosion overpressure inside silos (categories 1–3) has been developed considering the most important parameters governing the explosion process that are:

- Dust nature;
- Level of pre-ignition turbulence;
- Flame speed;
- Ratio l/d of enclosure;
- Vent coefficient;
- Vent activation overpressure.

This correlation is then extended to near cubical volumes of category 4.

The relevance of the proposed correlation is due to the following points:

- Harmonization of variables, without apparent link, as volume, shape, operating conditions (e.g. Level of turbulence), position of the ignition source;
- Capability of prediction of the overpressure developed in the course of an explosion;
- Capability of realistic vent sizing with regard to real work condition.

Relationship Definition

Propaedeutic considerations together with references to the work of other authors shall be done in order to establish a rational basis for the correlation adopted.

Turbulence production in the flow ahead of the flame front can significantly increase the rate of combustion in pipes and elongated enclosures. Fig. 15 illustrates how this mechanism leads to flame acceleration and pressure build-up.

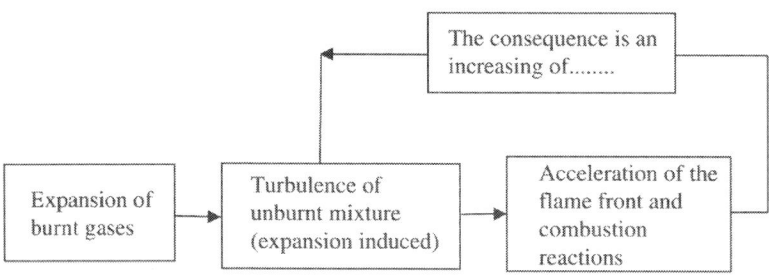

Figure 15: Positive feedback loop generating flame acceleration due to turbulence (inspired by Bjerktvedt et al., 1997).

A clear demonstration of this mechanism is shown in Fig. 16 that reports the evolution of the measured flame speed of methane–air mixture along a tube with inside diameter of 1.4 m and length of 40 m (Bartknecht, 1971). The ignition point was located at the closed end of the tube. The figure put in evidence the strong growth of the flame speed when the mixture is ignited at the closed end of the tube. In case of ignition at the open end, the burnt gases expand outside the tube and the unburnt mixture is not thrusted ahead. This results in a very low flame speed that quickly stabilizes on values of few metres per second. The combustion of gas–air mixtures inside enclosures has always been subject of interest in order to predict the effects of explosion both on structures and environment. The availability of empirical formulas based on experimental data could be useful either for prediction of the explosion overpressure or calculation of the vent size. The vent sizing of enclosures with high *L/D* ratio is substantially based on the work of Rasbash & Rogowski (1960), Rasbash & Rogowski (1961) and Rasbash & Rogowski (1963)performed on tubes of 75 and 150 mm diameter and ducts of 300 mm² section with length to diameter ratio close to L/D=50. Propane–air mixtures were used in the tests. Further experimental work was performed on square ducts with sides of 0.61 and 0.92 m and length to width ratios up to L/L_{side}=15 byTite, Binding, and Marshall (1991). Methane–air mixtures were used in the tests. Also the Health and Safety Laboratory (Pritchard, Allsopp, & Eaton, undated) has recently carried out several tests using stoichiometric propane–air and methane–air mixtures with vented cylindrical steel enclosures (D =1.0m and L/D up to 11.8) and ducts (2.50 m×2.50 m and L/L_{side} up to 12). All the tests were conducted at zero pre-ignition turbulence with different positions of the ignition source. The worst cases, in terms of explosion overpressure, were recorded when ignition source and vent panel were located on the opposite end of tubes or ducts. The dimensions of the enclosures used in the work of Pritchard, Allsopp, and Eaton (undated) are in such a way comparable with those of typical industrial equipments without mediation of any scale factor. All the experimental works cited above underline a direct relationship between flame front speed S_F (and hence overpressure) and length to diameter ratio (L/D) of the enclosures tested. Therefore, fundamental variables for the prediction of explosion overpressure assumed in this work are: length to diameter ratio L/D, flame speed S_F, vent coefficient K_{vent} and vent activation overpressure P_{vent} (Harris, 1983). The inertia of vent panels is here neglected assuming the vent covers to have a low inertia as for the experimental works considered in this paper. In addition a relevant variable has been introduced: the rms pre-ignition turbulence intensity u'_{rms}.

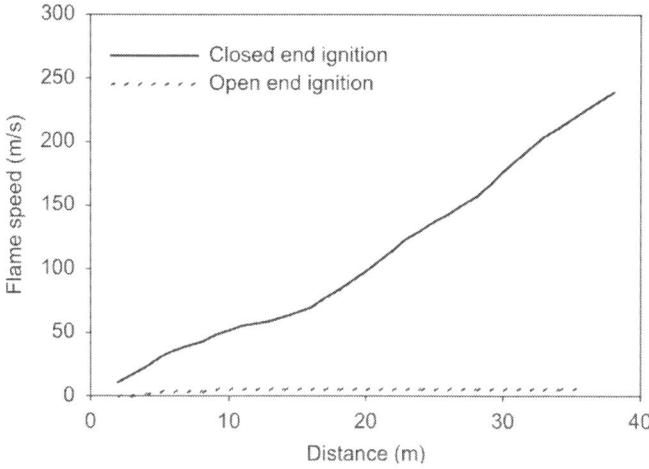

Figure 16: Flame front speed versus distance inside a tube of 1.4 m diameter and 40 m length. Methane–air mixture (Bartknecht, 1971).

It is just the case to remark that the effects of turbulence are taken into account by two points of view: the first relates to the pre-ignition turbulence, already working into the empirical correlation for flame speed presented previously; the second relates to the expansion generated turbulence accounted for by the length to diameter ratio of the enclosure and working as an enhancement factor of the flame speed. The relationship developed for prediction of the explosion overpressure inside silos, in case of bottom ignition, is as follows:

$$P_{ex} = P_{vent} + 6 S_F K_{vent} \left(\frac{L}{D}\right), \qquad (30)$$

where P_{ex} is the explosion overpressure, P_{vent} is the vent activation overpressure, $K_{vent} = A_{top}/A_{vent}$ is the vent coefficient, L/D is the length to diameter ratio of the silo and S_F is the speed of the flame front supplied by Eq. (27) with the expansion factor set equal to $E = E^0 \cong P_{max}$ (gauge) and absolute pressure to p=1 bara, conditions that well agree with the case of low overpressure typical of vented explosions. Eq. (27) adjusted as above stated, leads to Silvestrini (2004) already reported Eq. (30) that was calibrated over a limited amount of experimental data.

$$S_F = \frac{K_{St}}{(p_{max}/p_0)^{1/\gamma}} \left[\frac{0.11}{p_{max}^{0.14}} + 0.15 u'_{rms}\right]. \qquad (31)$$

Actual calibration has been operated on the basis of the experimental data taken from works of various researchers and provided in Table 5, Table 6, Table 7, Table 8 and Table 9.

Table 5: 9.4 m³ silo—(12 m³ partly loaded) (Hauert & Vogl, 1995; Hauert et al., 1996

Dust	Maize starch									
K_{St}	140									
P_{max} (bar g)	9									
u'_{rms}	2.1				0.80*					
Feeding system	Perforated dispersion ring VDI 3673/ ignition delay time: 0.85 s					Pneumatic pipeline injection (axial) from silo top/ignition during injection				
A_{top} (m²)	2.0	2.0	2.0	2.0	2.0	2.0	2.0	2.0	2.0	2.0
A_{vent} (m²)	0.3	0.5	0.5	0.75	0.15	0.15	0.5	0.5	0.3	0.3
P_{vent} (mbar g)	100	100	100	100	100	100	100	100	100	100
L/D	3	3	3	3	3	3	3	3	3	3
K_{vent}	6.7	4	4	2.68	13.4	13.4	4	4	6.7	6.7
P_{ex} meas (mbar g)	1525	1550	825	893	537	1550	1502	332	427	648	743

Table 6: 20 m³ Switzerland silo (Eckhoff, 1997)

Dust	Maize starch
K_{St}	226

P_{max} (bar g)	10.3	
u'_{rms}	0.80*	1.5
Feeding system	Pneumatic pipeline injection (axial) from silo top/ignition during injection (same turbulence level of the 12 m³ silo)	Perforated dispersion ring VDI 3673/ignition delay time: 0.9 s (Bartknecht, 1985). The turbulence intensity has been derived from that of the 12 m³ silo knowing that both silos have the same diameter (D=1.60 m) but different length of 5 and 10 m. Since the 12 m³ silo has four dispersion rings, the 20 m³ silo should have eight rings to produce the same turbulence at the same delay time, for the same silo diameter and the same dispersion system using pressurized bottles. Really the 20 m³ silo has only six rings and hence the turbulence can be reduced to 2.1*(6/8)≈1.5 m/s

A_{top} (m²)	2	2	2	2	2	2	2	2
A_{vent} (m²)	1.3	1.15	0.8	0.5	2	1.5	0.8	0.5
P_{vent} (mbar g)	100	100	100	100	100	100	100	100
L/D	6.25	6.25	6.25	6.25	6.25	6.25	6.25	6.25
K_{vent}	1.54	1.74	2.5	4	1	1.33	2.5	4
P_{ex} meas (mbar g)	500	723	900	1445	600	600	1250	1783

Table 7: 500 m³ Norwegian silo (Eckhoff, 1997; Eckhoff & Fuhre, 1984)

Dust	Wheat grain					Maize starch
K_{St}	88					115
P_{max} (bar g)	7					7.5
u_{rms}'	0					1.5* (hypothetical)
Feeding system	Pneumatic pipeline injection (axial) from silo top/ignition after some seconds to have a quiescent cloud					Pneumatic pipeline injection (axial) from the top of the silo/ignition during injection
A_{top} (m²)	23.07	23.07	23.07	23.07	23.07	23.07
A_{vent} (m²)	2.00	3.2	3.4	4.82	8.8	8.8
P_{vent} (mbar g)	10–20	10–20	10–20	10–20	10–20	10–20
L/D	4	4	4	4	4	4
K_{vent}	11.54	7.21	6.78	4.79	2.66	2.66
P_{ex} meas. (mbar g)	435	470	300	249	126	550 (turbulent jet from the top of the silo)

Table 8: 236 m³ Norwegian silo and 20 m³ Switzerland silo (Eckhoff, 1997)

	236 m³ Norwegian silo	20 m³ Switzerland silo
Dust	Maize starch	
K_{St}	226	
P_{max} (bar g)	10.3	
u_{rms}'	0.3 (supposed)	0.80*

Correlations for Flame Speed and Explosion Overpressure of ...

Feeding system	Pneumatic pipeline injection (axial) from the top of the silo/ignition after some seconds to have a quiescent cloud	Pneumatic pipeline injection (axial) from the top of the silo/ignition during injection
A_{top} (m²)	10.67	2
A_{vent} (m²)	3.4	0.5
P_{vent} (mbar g)	100	100
L/D	6	6.25
K_{vent}	3.05	4
P_{ex} meas. (mbar g)	770	1330

Table 9: Elongated test chamber $V=0.074$ m³ (Scheid et al., 2006)

Dust	Maize starch (10–20 μm)									
K_{St}	41									
P_{max} (bar g)	6.44									
u_{rms}'	0	0.225*	0	0.225*	0	0.225*	0.16*	0.16*	0.16	0.16*
Turbulence production system	16 rotating fans at four equidistant levels									
A_{top} (m²)	0.07									
A_{vent} (m²)	0.0013	0.0013	0.002	0.002	0.0039	0.0039	0.0013	0.002	0.0039	0.002
P_{vent} (mbar g)	145	145	125	125	70	70	145	125	70	125
L/D	3.4									
K_{vent}	54.4	54.4	35.3	35.3	18.12	18.12	54.4	35.3	18.12	35.3
P_{ex} meas. (mbar g)	866	1470	612	1000	280	577	1160	800	415	770

Fig. 17 shows the comparison between explosion overpressure measured and calculated by Eq. (30).

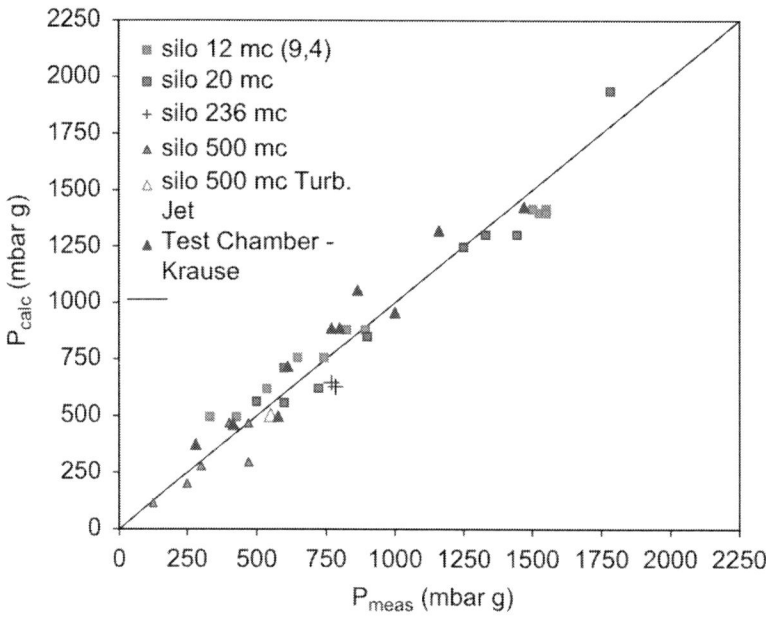

Figure 17: Comparison of explosion overpressure measured and calculated by Eq. (30)—the range of tolerance is: ±160mbr.

Turbulence intensities u'_{rms} listed in the previous Table 5, Table 6, Table 7, Table 8 and Table 9 and marked with (*) are half of the maximum value of turbulence recorded at the silo axis in the experimental works referenced above. The use of an averaged value of u'_{rms} has been assumed in those cases where the turbulence was mainly concentrated in the core of the silo. Fig. 18 shows a non-homogeneous turbulence distribution for three different cross sections of the 12 m³ silo. In all the other cases (e.g. use of perforated ring nozzles), no averaged values of turbulence have been used in Eq. (31), assuming a better homogeneity of turbulence intensity inside the enclosure. The averaged cross section turbulence intensity has also been applied to the data of experimental work of Scheid, Geissler, and Krause (2006) for the same reason stated above.

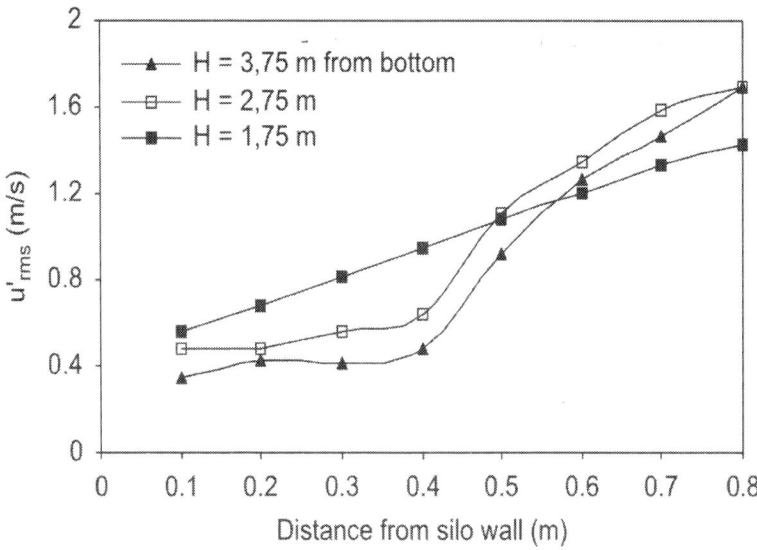

Figure 18: The rms turbulent velocity for three cross sections of the 12 m³ silo. Pneumatic pipeline injection (axial) from top of the silo top—feeding velocity 23 m/s (Hauert & Vogl, 1995).

About the level of turbulence intensity typical of industrial plant equipments a brief guidance is given in the following. Empirical formulas and characteristic values are reported in order to allow a rough estimate of turbulence intensity inside enclosure in realistic work conditions. Considering the 12 m³ silo (Hauert & Vogl, 1995), fed by an axial pneumatic pipeline injection system (pipeline diameter: =75 mm), and using the few available data, including the point at zero flow velocity corresponding to absence of turbulence, it is possible to obtain a simple relationship between rms turbulence intensity at the silo axis (1-m under the pipe outlet) and the flow velocity in the pipeline. Fig. 19 shows the increasing of turbulence with the velocity of the feeding flow for the 12 m³ silo.

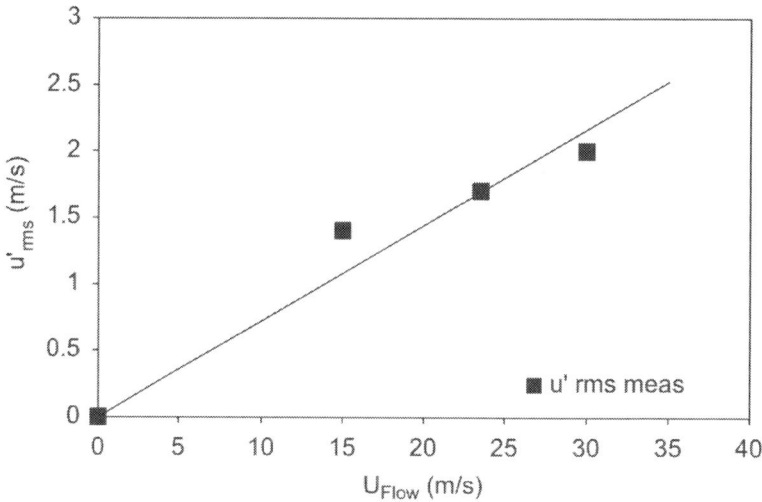

Figure 19: Turbulence intensity at the axis of the 12 m³ silo (1 m under the pipe outlet) (Hauert and Vogl, 1995; Hauert et al., 1996).

A linear proportionality between turbulence intensity and flow velocity can be assumed for practical use. On a nearly linear relationship is also based the following formula used in the computational code FLUENT: $u'_{rms} = 0.16 U_{Flow} \, Re^{-1/8}$, where the turbulent fluctuation at the pipe axis is a function of $U_{Flow}^{7/8}$.

The trend shown in Fig. 19 (turbulence at the silo axis) is given by the expression:

$$u'_{rms} = 0.072 U_{Flow} \tag{32}$$

and the average turbulence intensity in the whole silo, for the reasons explained above, can be considered about half of that at the silo axis:

$$\frac{d_{Pipeline}}{D_{Silo}} = \frac{75 \, mm}{1600 \, mm} = 0.047. \tag{33}$$

By means of a similitude with the 12 m³ silo, Eq. (33) applies, for a rough estimate of average turbulence intensity, to all those operating conditions similar to that shown in Fig. 20 where the ratio between the diameter of feeding pipeline and the diameter of silo is close to the following value (12 m³ silo):

$$\frac{d_{Pipeline}}{D_{Silo}} = \frac{75 \text{ mm}}{1600 \text{ mm}} = 0.047.$$

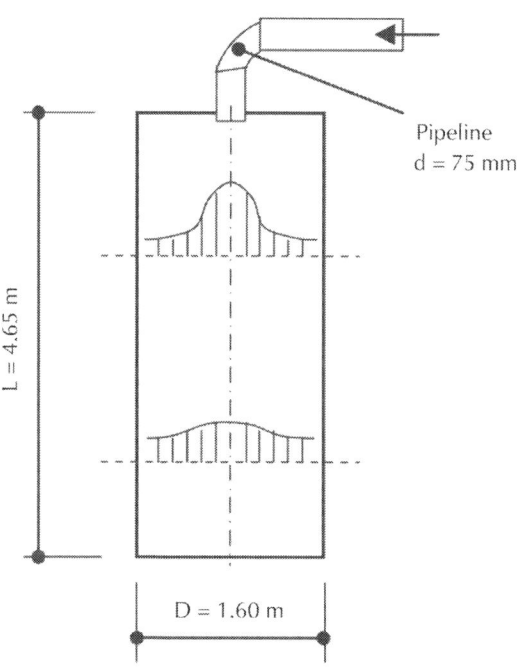

Figure 20: Qualitative behaviour of the rms turbulence intensity inside the 12 m³ silo fed by an axial pneumatic pipeline placed on the silo roof.

Hauert and Vogl (1995) and Hauert, Vogl, and Radandt (1996) performed measurements of turbulence intensity inside the 12 m³ silo, using both axial and tangential pneumatic feeding systems. They also used a mechanical filling system (screw conveyor) placed on the top of the silo. For pneumatic tangential feeding, at maximum flow velocity, a maximum turbulence level of $u'_{rms} = 0.4\,\text{m/s}$ was measured, while using the screw conveyor the maximum turbulence measured was $u'_{rms} = 0.5\,\text{m/s}$. For a better scenario depicting, measured values of turbulence decay inside typical enclosures are given here below:

- 1.3 m³ Tamanini sphere (isotropic value) (Tamanini, 1998):

$$u'_{rms} = 1.286(t - 0.34)^{-0.803} \text{ m/s for } t \geqslant 0.4\,\text{s},$$

- 64 m³ Tamanini room equipped with 16 pepper pot nozzles placed on opposite side wall (isotropic value) (Tamanini, 1998):

$$u'_{rms} = 21.87 - 17.30t \text{ m/s for } t \leqslant 0.73\,\text{s},$$

$$u'_{rms} = 3.72 t^{-2.78} \text{ m/s for } t > 0.73\,\text{s},$$

- 12 m³ silo equipped with perforated ring nozzles placed at four different heights (isotropic value) (Hauert et al., 1996):

$$u'_{rms} = 7.70\, e^{-2.16 t} + 0.85 \text{ m/s}.$$

For completeness mention to the work of Siwek et al. (2004) on conventional spray dryers must be done. Values of turbulence intensity (rms) measured in the lower part of four different dryers, ranging from 6 to 310 m³, was found to lie in the range $1 \leq u'_{rms} \leq 2\,\text{m/s}$.

The last topic treated in this paper regards the extension of the correlation found for the explosion overpressure prediction in elongated enclosures to enclosures with low (L/D). This requires the following propaedeutic consideration. The empirical correlation for explosion overpressure given by Eq. (30) relates to the case of bottom ignition. If the ignition source is located in proximity of half the silo length Eq. (30) can be used with half-coefficient on the form:

$$P_{ex} = P_{vent} + 3S_F K_{vent}\left(\frac{L}{D}\right). \tag{34}$$

In fact this condition produces experimental values of explosion overpressure that are about half of those obtained in the case of bottom ignition.

The same Eq. (34) can be succesfully applied to the case of near cubical volumes with central ignition using L/D≈1. Fig. 21 shows a comparison between the explosion overpressure measured (Tamanini (1990) and Tamanini (1998)) and calculated by Eq. (34). The turbulence intensity used for this calculation is that measured by Tamanini without any modification because it was probably quite homogeneous in the test volume, thanks to the dispersion method used. In fact the dispersion system consisted of 16 nozzles (like pepper pot), placed on two opposite side wall of the room and fed by pressure vessels at 8.3 bar. The effect is comparable to that of the dispersion rings.

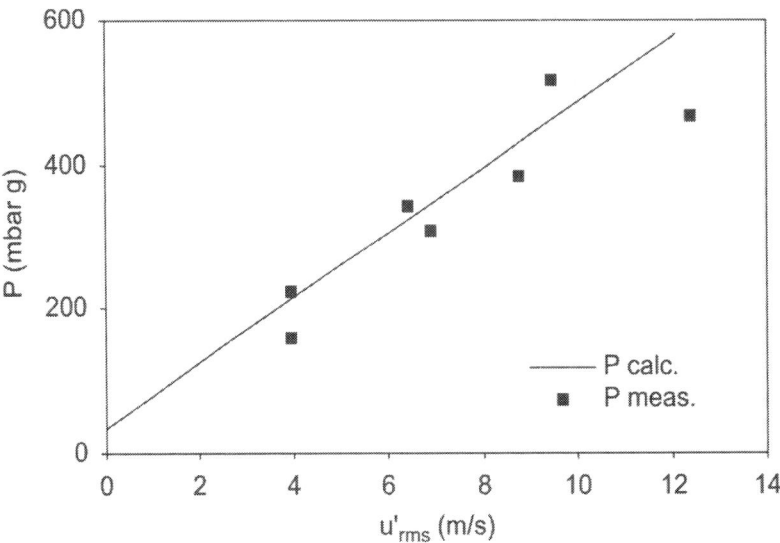

Figure 21: Comparison between explosion overpressure measured (64 m³ Tamanini room) and calculated by Eq. (34) for near cubical volumes (L/D=1).

At the end is just the case to remind that the method proposed for the evaluation of the explosion overpressure inside industrial enclosures is only a simplified method and more accurate evaluations can today be done by means of dedicated computational fluid dynamic codes like DESC (Skjold, 2007). Hand calculations like those here presented can be used only in simple cases and become no longer suitable for complex geometries and flow fields.

Fig. 22 shows the vertical component of feeding velocity and the rms turbulence intensity in the 12 m³ silo (Hauert et al., 1996) obtained with the CFD code FLACS (Silvestrini, Genova, & Leon, 2007). The simulation refers to the case of pneumatic pipeline injection (axial) at feeding velocity of 23 m/s. The agreement between measured (Fig. 18) and calculated (Fig. 22) values of u'_{rms} is very satisfactory.

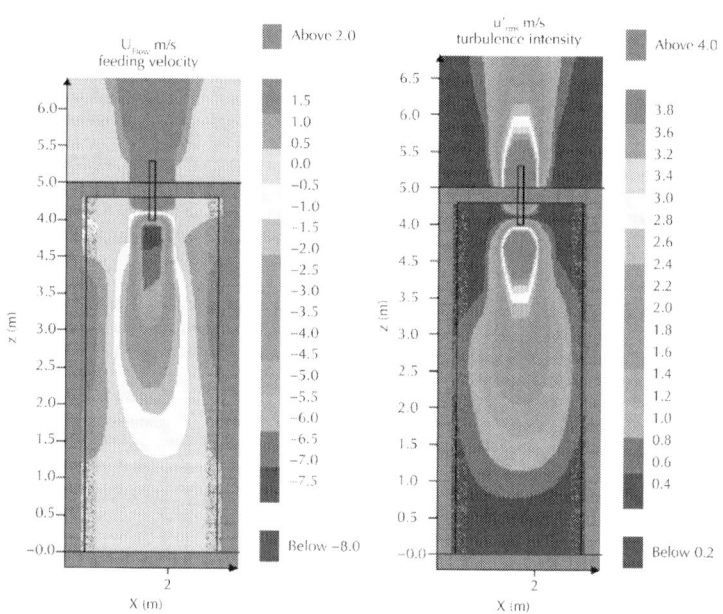

Figure 22: Vertical component of flow velocity and rms turbulence intensity for the 12 m³ silo equipped with axial pneumatic pipeline working at feeding velocity of 23 m/s. The simulation has been made with the FLACS code (Silvestrini, Genova, & Leon, 2007).

CONCLUSIONS

Explosion relief vents on enclosures in powder-handling plants are currently designed according to technical standards that in some situations may overestimate the required vent area significantly but also to underestimate it in case of strong turbulent flow operating conditions. In fact the real work conditions of industrial plants (e.g. actual turbulence level) are not always considered. The availability of methods either for evaluation of the explosion overpressure or calculation of the vent size involving the pre-ignition turbulence could be very useful for a more realistic estimate of these quantities. In this work two empirical correlations are presented: the first one allows the calculation of the flame speed and the burning velocity of a dust–air mixture by explosion indices K_{St} and P_{max} of the standardized 20-l sphere. The second allows the calculation of the explosion overpressure inside an enclosure also in case of turbulent dust clouds.

REFERENCES

1. Andrews, G. E., & Bradley, D. (1972). Determination of burning velocities: a critical review. Combustion and Flame, 18, 133–153.
2. Ballal, D. R. (1983). Flame propagation through dust clouds of carbon, coal, aluminium and magnesium in an environment of zero gravity. Proceedings of the Royal Society of London, A385, 21–51.
3. Bartknecht, W. (1971). Brenngas und Staubexplosionen Forschungsbericht F45. Koblenz, W. Germany: Bundesinstitut fur Arbeitsschutz (Bifa).
4. Bartknecht, W. (1981). Explosions. Course, prevention, protection. Berlin: Springer.
5. Bartknecht, W. (1985). Effectiveness of explosion venting as a protective measure for Silos. Plant/Operations Progress, 4(1).

6. Bjerktvedt, D., Bakke, J. R., & van Wingerden, K. (1997). Gas explosion handbook. Journal of Hazardous Materials, 52, 1–150.
7. Botha, J. P., & Spalding, D. B. (1954). The laminar flame speed of propane/air mixtures with heat extraction from the flame. Proceedings of the Royal Society of London, A225, 71–96.
8. Bradley, D., Chen, Z., & Swithenbank, J. R. (1988). Burning rates in turbulent fine dust–air explosions. Proceedings of the 22nd international symposium on combustion (pp. 1767–1775).
9. Bradley, D., & Lee, J. H. S. (1984). Proceedings of the first international colloquium on the explosibility of industrial dusts (Vol. 2, pp. 220–223).
10. Cassel, H. M. (1964). Some fundamental aspects of dusts flames. Report Inv. 6551, US Bureau of Mines, Washington.
11. Dahoe, A. E., Cant, R. S., & Scarlett, B. (2001). On the decay of turbulence in the 20-l explosion sphere. Flow, turbulence and combustion (pp. 159–184). Dordrecht: Kluwer Academic Publishers.
12. Dahoe, A. E., & de Goey, L. P. H. (2003). On the determination of the laminar burning velocity from closed vessel gas explosions. Journal of Loss Prevention in the Process Industries, 16, 457–478.
13. Dahoe, A. E., Hanjalic, K., & Scarlett, B. (2002). Determination of the laminar burning velocity and the Markstein length of powder–air flames. Powder Technology, 122, 222–238.
14. Dahoe, A. E., Zevenbergen, J. F., Lemkowitz, S. M., & Scarlett, B. (1996). Dust explosions in spherical vessels: The role of flame thickness in the validity of the "cube-root law". Journal of Loss Prevention in Process Industries, 9, 33–44.
15. Drysdale, D. (1998). An introduction to fire dynamics (2nd ed.). Chichester, England: Wiley.
16. Eckhoff, R. K. (1997). Dust explosions in the process industries (2nd ed.). Oxford: Butterworth-Heineman.

17. Eckhoff, R. K., & Fuhre, K. (1984). Dust explosion experiments in a vented 500 m3 silo cell. Journal of Occupational Accidents, 6, 229–246.
18. Gieras, M., Glinka, W., Klemens, R., & Wolanski, P. (1995). Investigation of flame structure during laminar and turbulent burning in dust–air mixtures. Conference on dust explosions. Protecting people, equipment, buildings and environment (pp. 168–224).
19. Gieras, M., Klemens, R., & Wolanski, P. (1996). Evaluation of turbulent burning velocity for dust mixtures. Proceedings of the seventh international colloquium on dust explosions (pp. 535–551), Bergen, Norway.
20. Glinka, W., Wang, X., Wolanski, P., & Xie, L. (1996). Velocity and structure of laminar dust flames. Proceedings of the seventh international colloquium on dust explosions (pp. 61–68), Bergen, Norway.
21. Goroshin, S., Mamen, J., Lee, J., & Sacksteder, K. (2005). Ground-based and microgravity study of flame quenching distance in metal dust suspension. In Proceedings of the 20th international colloquium on the dynamics of explosions and reactive systems, McGill University, Montreal, Canada.
22. Han, O. S., Yashima, M., Matsuda, T., Matsui, H., Miyake, A., & Ogawa, T. (2000). Behaviour of flames propagating through Lycopodium dust clouds in a vertical duct. Journal of Loss Prevention in the Process Industries, 13, 449–457.
23. Harris, R. J. (1983). The investigation and control of gas explosions in buildings and heating plants. London, New York: E & FN Spoon Ltd.
24. Hattwig, M., & Steen, H. (2004). Handbook of explosion prevention and protection. Weinheim: Wiley-VCH Verlag GmbH & Co.
25. Hauert, F., & Vogl, A. (1995). Measurement of dust cloud characteristics in industrial plants. Final technical report: Protecting people, equipment, buildings and environment against dust explosions—CREDIT Project.

26. Hauert, F., Vogl, A., & Radandt, S. (1996). Dust cloud characterization and its influence on the pressure–time history in silos. Process Safety Progress, 15, 178–184.
27. Horton, M. D., Goodson, F. P., & Smoot, L. D. (1977). Characteristics of flat, laminar coal dust flames. Combustion and Flame, 28, 187–195.
28. Kaesche–Krischer, B. (1959). Untersuchungen an Vorgemischte, Laminairen Staub/Luft-Flammen. Staub, 19, 200–203. Kolbe, M. (2001). Laminar burning velocity measurements of stabilized aluminium dust flames. Thesis, Department of Mechanical Engineering, Concordia University, Montreal, Quebec, Canada.
29. Krause, U., & Kasch, T. (2000). The influence of flow and turbulence on flame propagation through dust–air mixtures. Journal of Loss Prevention in the Process Industries, 13, 291–298. Krause, U., Kasch, T., & Gebauer, B. (1996). Velocity and concentration effects on the laminar burning velocity of dust–air mixtures. Proceedings of the seventh international colloquium on dust explosions, Bergen, Norway (pp. 51–54).
30. Lewis, B. (1954). Selected combustion problems. AGARD (Butterworths) p. 177.
31. Lewis, B., & von Elbe, G. (1987). Combustion, flames and explosions of gases (3rd ed.). Orlando, FL: Academic Press.
32. Mazurkiewicz, J., & Jarosinski, J. (1990). Temperature and laminar burning velocity of Cornstarh dust–air flames. Proceedings of the fourth international colloquium on dust explosions. Poland: PorabkaKozubnik.
33. Mazurkiewicz, J., & Jarosinski, J. (1991). Investigation of burning properties of cornstarch dust–air flames. In Grain dust explosions and control, Warschau (pp. 63–90).
34. Mazurkiewicz, J., & Jarosinski, J. (1994). Investigation of a laminar cornstarch dust–air flame front. In Proceedings of the sixth international colloquium on dust explosions (pp. 179–185). Shenyang: Northeastern University Press.

35. Metghalchi, M., & Keck, J. C. (1982). Burning velocities of mixtures of air with methanol isooctane, and indolene at high pressure and temperature. Combustion and Flame, 48, 191–210.
36. Milton, B. E., & Keck, J. C. (1984). Laminar burning velocities in stoichiometric hydrogen and hydrogen–hydrocarbon gas mixtures. Combustion and Flame, 58, 13–22.
37. NFPA 68 (1994). Guides for venting of deflagration, National Fire Protection Association, Quincy, MA.
38. Pedersen, L. S., & van Wingerden, K. (1995). Measurement of fundamental burning velocity of dust–air mixtures in industrial situations. In Dust explosions, protecting people, equipment, buildings and environment (pp. 140–167). British Materials Handling Board.
39. Peters, N. (1992). Fifteen lectures on laminar and turbulent combustion. Aachen, Germany: Ercoftac Summer School.
40. Pritchard, D. K., Allsopp, J. A., & Eaton, G. T. (undated). Gas explosion venting in elongated enclosures. Paper presented at /www. safetynet.deS. ARTICLE IN PRESS M. Silvestrini et al. / Journal of Loss Prevention in the Process Industries 21 (2008) 374–392 391
41. Proust, C. (1993). Experimental determination of the maximum flame temperatures and of the laminar burning velocities for some combustible dust–air mixtures. In Proceedings of the fifth international colloquium on dust explosions, Pultusk, Poland (pp. 161–175).
42. Proust, C. (2006). Flame propagation and combustion in some dust–air mixtures. Journal of Loss Prevention in the Process Industries, 19, 89–100.
43. Proust, C., Roux, P., & Chhuon, B. (2000). Pre´voir les Effects des Explosions de Poussie`res sur l'Environnement. EFFEX, un Outil de Simulation. INERIS/DRA 22751 (pp. 1–50).
44. Proust, C., & Veyssiere, B. (1988). Fundamental properties of flames propagating in starch dust–air mixtures. Combustion Science and Technology, 62, 149–172.

45. Pu, Y. K., Jarosinski, J., Johnson, V. G., & Kauffman, C. W. (1990). Turbulence effects on dust explosion in the 20-l spherical vessel. In Proceedings of the 23rd international symposium on combustion (pp. 843–849). Pittsburgh: The Combustion Institute.
46. Pu, Y. K., Jarosinski, J., Tai, C. S., Kauffman, C. W., & Sichel, M. (1988). The investigation of the feature of dispersion induced turbulence and its effects on dust explosions in closed vessels. In Proceedings of the 22nd international symposium on combustion, The Combustion Institute, Pittsburgh (pp. 1777–1787).
47. Pu, Y. K., Jia, F., Wang, S. F., & Skjold, T. (2007). Determination of the maximum effective burning velocity of dust–air mixtures in constant volume combustion. Journal of Loss Prevention in the Process Industries, 20, 462–469.
48. Pu, Y. K., Lee, Y. C., Kauffman, C. W., & Bernal, L. P. (1989). Determination of turbulence parameters in closed explosion vessel. In 12th international colloquium on the dynamics of explosions and reactive systems, Ann Arbor, MI (pp. 107–123).
49. Rasbash, D. J., & Rogowski, Z. W. (1960). Relief of explosions in duct systems. In Proceedings of the first symposium on chemical process hazards (pp. 58–65). England, London: Institution of Chemical Engineers.
50. Rasbash, D. J., & Rogowski, Z. W. (1961). Gaseous explosions in vented ducts. Combustion and Flame, 4, 301–312.
51. Rasbash, D. J., & Rogowski, Z. W. (1963). Relief of explosions in propane-air mixtures moving in a straight unobstructed duct. In Proceedings of the second symposium on chemical process hazards with special reference to plant design. England, London: Institution of Chemical Engineers.
52. Scheid, M., Geissler, A., & Krause, U. (2006). Experiments on the influence of pre-ignition turbulence on vented gas and dust explosions. Journal of Loss Prevention in the Process Industries, 19, 194–199.

53. Silvestrini, M. (2004). Esplosioni di Polveri nell'Industria: Sorgenti d'Ignizione - Combustione Turbolenta. Tesi, Scuola di Specializzazione in "Sicurezza e Protezione," Universita` degli Studi di Roma "La Sapienza," Italy, Rome.
54. Siwek, R. (1988). Reliable determination of safety characteristics in the 20- litre apparatus. Proceedings of Conference on Flammable Dust Explosions, November 2–4, St. Louis, Missouri.
55. Siwek, R., van Wingerden, K., Hansen, O. R., Sutter, G., Kubainsky, Chr., Schwartzbach, Chr., Giger, G., & Meili, R. (2004). Dust explosion venting and suppression of conventional spray dryers. From the final research report "Explosion Protection of Conventional Spray Dryers" (Van Wingerden, K. and Siwek, R., Final Research Report 2004).
56. Skjold, T. (2003). Selected aspects of turbulence and combustion in 20-l explosion vessels. Thesis, Department of Physics of University of Bergen, Norway.
57. Skjold, T. (2007). Review of the DESC project. Journal of Loss Prevention in the Process Industries, 20, 291–302.
58. Tai, C. S., Kauffman, C. W., & Sichel, M. (1988). Turbulent dust combustion in a jet-stirred reactor. Progress in Astronautics and Aeronautics, 113, 62–86.
59. Tamanini, F. (1990). Turbulence effects on dust explosion venting. Plant/ Operation Progress, 9, 52–60.
60. Tamanini, F. (1998). The role of turbulence in dust explosions. Journal of Loss Prevention in the Process Industries, 11, 1–10.
61. Tezok, F. I., Kauffman, C. W., Sichel, M., & Nichols, J. A. (1986). Turbulent burning velocity measurements for dust–air mixtures in a constant volume spherical bomb. Progress in Astronautics and Aeronautics, 105, 184–195.
62. Tite, J. P., Binding, T. M., & Marshall, M. R. (1991). Explosion relief for long vessels. Paper presented at the conference on fire and explosion hazards, Moreton-in-Marsch, April. Van der Wel, P. (1993). Ignition and propagation of dust explosions. Ph.D. Thesis, Delft University Press.

63. Van Wingerden, K. (1996). Simulation of dust explosions using a CFDCode. Proceedings of the seventh international colloquium on dust explosion, Bergen, Norway (pp. 642–651).
64. VDI 3673. (2002). Pressure venting of dust explosions. Verein Deutscher Ingenieure.
65. Veynante, D., & Vervisch, L. (2002). Turbulent combustion modelling. Progress in Energy and Combustion Science, 28, 193–266.
66. Wang, S., Pu, Y., Jia, F., Gutkowski, A., & Jarosinski, J. (2006). An experimental study on flame propagation in cornstarch dust clouds. Combustion Science and Technology, 178, 1957–1975.
67. Zalosh, R. (2007). New dust explosion venting design requirements for turbulent operating conditions. Journal of Loss Prevention in the Process Industries, 20, 530–535.
68. Zevenbergen, J. F. (2004). Turbulent burning velocity measurements in 20-l sphere. DESC project. The Explosion Group of Delft University of Technology.

Chapter 7

Dust Explosions: CFD Modeling as a Tool to Characterize the Relevant Parameters of the Dust Dispersion

Carlos Murillo[a], Olivier Dufaud[a], Nathalie Bardin-Monnier[a], Omar López[b], Felipe Munoz[c], Laurent Perrin[a]

[a]Reaction and Chemical Engineering Laboratory (LRGP), University of Lorraine, UPR 33 49 CNRS, 1 Rue Grandville, BP 20451, 54001 Nancy, France

[b]Mechanical Engineering Department, Universidad de los Andes, Carrera 1 Este 19A 40, Bogotá, Colombia

[c]Chemical Engineering Department, Universidad de los Andes, Carrera 1 Este 19A 40, Bogotá, Colombia

ABSTRACT

This study proposes two different approaches to describe the dynamics of a transient gas–solid flow inside a modified Hartmann tube or a similar apparatus. An experimental analysis was performed to analyze the main influencing parameters during a typical characterization test of combustible solids. For this purpose, the development of dust clouds of different solid materials inside the tube has been evidenced by high-speed videos and particle-size measurements. The two-phase flow induces variations in particle size distributions and segregation levels that have been identified during the dispersion process.

Furthermore, a computational fluid dynamics (CFD) simulation, based on an Euler–Lagrange approach, has been developed with ANSYS FLUENT™. It was directed towards the assessment of flow conditions pertaining to the agglomeration and fragmentation of particles dispersed in the standard setup. The experimental data have been used to adjust a fragmentation model in the CFD simulation in order to describe the variations in particle size distributions accurately.

These approaches have evidenced three different stages of solids dispersions, which can be classified according to variations in particle size distributions: a fragmentation phase, stabilization of the dust cloud and sedimentation phase. These facts suggest that, for the micrometric aluminum powders which have been tested, the minimum ignition delay must be 60 ms, after considering the distribution of the discrete phase inside the tube and the variations in size distribution at the elevation of ignition electrodes. This delay being greatly influenced by powders properties, this paper suggests combining both simulation and experimental approaches in order to determine the most relevant or the most conservative conditions for the determination of dust flammability.

INTRODUCTION

Combustion of solid materials, especially dust explosions, is particularly influenced by the dispersion characteristics of the particles in the combustion air. As highlighted by Eckhoff (2009), the most important properties of the dust dispersion are: (i) the particle shape, (ii) the particle size distribution (PSD), (iii) the agglomeration degree, (iv) the dust concentration within the cloud and (v) the degree of turbulence of the suspension. The influence of the last four parameters will be discussed in this article.

The effect of the particle size distribution on dust explosion has already been extensively studied (Eckhoff, 1996). It is generally considered that, by reducing the particle size, the dust ignitability increases as well as its explosivity (Callé et al., 2005 and Matsuda et al., 2001); a modification of the powder diameter having notably a strong influence on the rate-limiting step of the oxidation and on the persistence of the cloud. However, the expression "particle size" is too imprecise and appropriate particle size characteristics should be chosen in order to represent such phenomena. If the mean volume diameter is often chosen for the sake of convenience and for industrial applications, it is not always the best indicator in terms of reactivity. Indeed, dust ignitions have been strongly linked to the presence of small particles, due for instance to the presence of powder attrition or erosion (Amyotte et al., 2009) and the entire particle size distribution brings relevant information, especially the analysis of the fine tail of the distribution (Eckhoff, 2009). This remark is even more accurate when dealing with nanoparticles.

The degree of agglomeration of the powder, defined as the ratio between the collision diameter of the agglomerates and the diameter of the primary particles, plays also a significant role in the dust explosion phenomenon. On the one hand, if the agglomerates are not broken by the dispersion process, they tend to behave as a large single particle of the size of the agglomerate (Eckhoff, 2003), burning with the same combustion regime. On the other hand, for dispersion processes of higher intensities, the particle size distribution of the pure powder does not correspond to the characteristics of the dust suspension. For very small particles (often below 10 μm) and especially for nanopowders, interparticle attractive forces such as Van der Waals forces cannot be neglected and the agglomerate cohesive strength impacts the dust ignition. For instance, Bouillard et al. (2010) have established that the Minimum Ignition Temperature (MIT) of carbonaceous particles suspensions decreases down to 670 °C when their BET (Brunauer–Emmett–Teller) diameters decreases from 110 to 40 nm, but increases for particles having primary particle sizes of 23 and 3 nm. This increase can be explained by the strong agglomeration level of those nanopowders, having respectively agglomerates diameters of 23 and 10 μm. In this specific case, the agglomeration phenomenon tends to prevent ignition.

A totally different behavior has been proposed by Trunov (2006) in the case of aluminum nanoparticles, illustrating also the influence of the agglomeration level on dust ignition. The author estimates that nano-size aluminum particle agglomerates can be easily ignited in comparison with single particles. Indeed, convective heat losses decrease for particle agglomerates, which promotes their self-heating.

In order to oxidize agglomerated powders, the amount of energy corresponding to the agglomerate cohesive strength should then be added to the Minimum Ignition Energy (MIE) of the primary particles. Erlend and Eckhoff (2006) also propose to consider a global dispersibility parameter related to the minimum work needed to break the interparticle bonds in order to quantify the influence of the agglomeration degree.

In addition, turbulence can affect the properties of the initial dust cloud and the flame propagation. In this work, we will ignore the explosion-induced turbulence to focus only on the turbulence of the initial dust cloud. Turbulence can both have a promoting effect on the dust explosivity (Eckhoff, 1977) or a quenching effect on the flame kernel growth (Glarner, 1984). It has obviously an impact on the mixing of fuel and oxidizer, on the efficiency of the heat transfer, but also on the dust concentration distribution in the cloud as well as on the agglomeration degree of the powder.

All these considerations must then been taken into account when measuring safety parameters related to dust explosion. During the twentieth century, several standardization techniques have been developed in order to experimentally determine the flammability and explosivity of combustible powders. Such parameters describe the explosive behavior of the mixture and also provide useful information for facilities that handle this type of materials. Among these methods, the modified Hartmann tube and the 20 l sphere constitute the most frequently used equipment for determining quantitative parameters such as the Minimum Ignition Energy (MIE) (IEC 1241-2-3 Ed. 1.0 b, 1994) and the Minimum Explosive Concentration (MEC). The 20 l sphere also provides pressure profiles that characterize the explosion severity of a combustible cloud composed by the gas and the solid disperse phase (ISO 6184-1, 1985). However, it appears more and more clearly that the tests conditions set by these standards do not systematically lead to the determination of the most representative, or even the safest, parameters with regard to the industrial context. For instance, the ignition delay (tv) set at 60 ms, which is the delay between the dust dispersion into the 20 l sphere and its ignition, is inadequate for most of the powders and even more for nanoparticles. As previously said, the link between the flame propagation and the suspension hydrodynamics constitutes a better knowledge of the initial suspension characteristics, which, enables an informed choice of the tests conditions.

The aim of this work is then to study the particles suspension in a turbulent flow, whose characteristics will match those encountered in our standardized equipment as for instance the explosion sphere, the modified Hartmann tube or an explosion tube designed for the study of flame propagation (Sanchirico et al., 2011). Some studies have been carried out in order to simulate, thanks to a 3D CFD model, the dust dispersion inside the 20 l sphere. Recently, Di Sarli et al. (2013) and Di Benedetto et al. (2013) have shown that the dust is not homogeneously dispersed within the vessel, due to the presence of multiple turbulent vortex structures. In the case of large particles, with diameters greater than 100 µm, the dust is mainly pushed towards the walls, which is qualitatively in accordance with the experimental results obtained by Kalejaiye et al. (2010) with optical dust probes. This article was focused on the explosion tube and two complementary approaches have been used. On the one hand, Computational Fluid Dynamics (CFD) simulations have been carried out to study the hydrodynamics of such suspensions. The two-phase flow simulation was based on an Euler–Lagrange approach because the amount of solid phase does not represent a high volume fraction in the mixture. With regard to the high solid loading, it is compulsory to characterize the particle–particle interactions and the potential fragmentation or agglomeration, but also to take the action of the particles upon the fluid into account. On the other hand, in situ particle size measurements and high-speed video cameras have been used to validate the model by identifying the particle size distributions, the agglomeration degree and the dust concentration within the experimental setup as a function of time. This paper presents an analysis that has been developed at micrometric and sub-micrometric scales to evaluate the behavior of the disperse phase in air by simulating the conditions inside the specific experimental apparatus. The latter will be briefly described as well as the powders used in this study. The CFD model, its assumptions, experimental validation, limitations and accuracy associated with the scale analyzed will then be developed.

EXPERIMENTAL CHARACTERIZATION OF DUST DISPERSION

Dusts Characteristics

Tests have been performed on various powders in order to validate the simulation models. In this paper, the results obtained for glass beads and aluminum powders will be presented. Glass beads have been chosen, even if they are obviously non-combustible, due to the fact that they are non-cohesive particles without agglomerates. In addition, various microsized and nanosized aluminum powders have been tested. Results obtained with aluminum nanopowders (200 nm) will not be presented here. Scanning electron microscopy has been performed on the aluminum samples. For micro-Al particles, the particles shapes are rather ellipsoidal than spherical, with a smooth surface. The main characteristics of the particle size distributions are given in Table 1.

Table 1: Main characteristics of the particle size distributions of the powders

Powders	Dispersion medium	d_{10} (µm)	d_{50} (µm)	d_{90} (µm)	$d_{3,2}$ (µm)
Glass beads	Ethanol	64	86	115	84
	Dry	56	75	84	39
Micro-Al 7	Ethanol	3	7	13	3
	Dry	1	7	15	3
Micro-Al 42	Ethanol	13	42	109	27
	Dry	7	34	77	17

Particle size measurements have been performed on wet samples with a laser diffraction analyzer (Mastersizer, Malvern Instrument) and on dried samples by using a laser diffraction sensor HELOS-VARIO/KR (SYMPATEC GmbH). In the latter case, the dusts have been introduced at the top of a glass tube and the particle size distributions have been measured during the sedimentation phase. By comparing the distributions, it can be seen that the characteristic diameters are significantly different for measurements performed in ethanol or on dried powders, except for the aluminum micrometric particles with a mean diameter of 7 µm.

Indeed, by changing the dispersion medium and technique, particles interactions such as Van der Waals attractive forces or electrostatic forces are strongly modified as well as the continuous medium viscosity; which notably implies that the viscous dissipation is altered from liquid phase to gas phase. These changes lead to modifications in the intensities and extent of aggregation or agglomeration phenomena, and can then explain the variations in the particle size distributions. As previously said, for small particles such as Micro-Al 7 (Table 1), the agglomerate cohesive strength is great, which enhances their stability. These facts also show that the particle size distributions obtained in water or a solvent, which are often used in dust explosion studies, do not fully correspond to the characteristics of the powders dispersed in air.

Experimental Set-Up

In order to characterize the ignition sensitivity of a powder, various parameters must be determined such as the minimum ignition temperature, energy and the minimum explosive concentration (MIT, MIE and MEC). The latter two parameters can be measured by using a modified Hartmann tube, by exposing the dusts clouds to an electrical spark whose energy ranges from 1 to 1000 mJ (IEC 1241-2-3 Ed. 1.0 b, 1994). This apparatus consists of a cylindrical glass tube with a diameter of 7 cm and a volume of 1.2 l. Powder samples are initially placed at the bottom of the tube and are then dispersed by an air blast of 7 bar. The delay between the beginning of the dust dispersion and the ignition is called ignition delay (tv); it is directly related to the turbulence level of the dust–air mixture. Generally, tv ranges between 60 and 180 ms but can be defined on a wider range. For this study and for performance of granulometric analyses, the cylindrical tube has been replaced by a vertical tube of 1 m height with a square cross section of 0.07×0.07 m. The bottom of the tube corresponds to a toroid with a diameter of 34 mm that is used to place the powder that will be characterized. A mushroom-shaped nozzle is located in this cup in order to disperse the dust homogenously (IEC 1241-2-3 Ed. 1.0 b, 1994). The tube has two opposite walls made of glass and two opposite walls made of stainless steel (Sanchirico et al., 2011). The injection system of the modified Hartmann tube has been reproduced with the same mushroom-shaped nozzle and the same dispersion conditions. The electrodes, which have been removed for this study, are usually located at 12.5 cm from the bottom of the setup. Specific attention has been paid on the dust cloud characteristics at this height.

A laser diffraction sensor HELOS-VARIO/KR (SYMPATEC GmbH) with an adjustable stand has been chosen to determine the particle size distribution during the transient dispersion process (Fig. 1). This apparatus has an optic system composed by a laser emission and a multi-element photo-detector with auto-alignment for optimum acquisition of the diffraction patterns. A characterization of particle size distributions can be performed by laser scattering with 32 different detectors located in a circular arrangement. The incident beam is generated by a 5 mW helium–neon source. For micrometric particles, a R3-lens has been used with the following measuring range: 0.5/0.9–175 µm; whereas a submicron module R1 (0.10/0.18–35 µm) has been used in the case of nanoparticles. The acquisition frequency can be set up to 2 distributions par millisecond. The Fraunhofer's extended theory has been considered for PSD determination. This approach considers opaque and non-porous particles. The particles dispersed in the dust cloud possess a random movement. These assumptions imply that the laser scattering is proportional to the particle size.

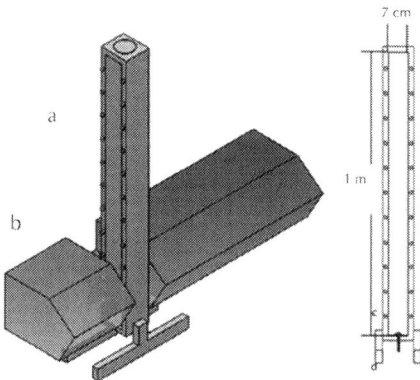

Figure 1: Experimental setup developed for the characterization of solids dispersion. (a) Dispersion tube, (b) Particle size distribution analyzer, (c) Dispersion nozzle, (d) Gas inlet.

A Phantom V91 high speed video-camera has also been used to record the dispersion of the dust clouds with frame rates ranging from 1000 to 5000 frames per second. As a consequence, this experimental setup has been considered for characterizing the dust cloud dynamics by determination of the particle size distributions and the dust concentrations during the dispersion process.

DESCRIPTION OF GAS–SOLID FLOWS BY SIMULATIONS BASED ON COMPUTATIONAL FLUID DYNAMICS

CFD Approaches for Gas–Solid Flow Dispersions

The description of gas–solid flows has been developed by different approaches according to the materials analyzed in the study case. Nevertheless, most of the numerical studies are based on Euler–Lagrange approach. For instance, Jenkins et al. (2013) studied small metallic particles (Al or W) driven by the product gas of an explosive and determined the drag coefficients and velocity fields combining Eulerian and Lagragian solvers. Jian et al. (2009) modeled propagation of potato starch dust explosions and obtained a very good representation of the acceleration of flame propagation. Kosinski (2011) analyzed the first moment of explosion suppression by a cloud of particles. Bidabadi et al. (2010) modeled the velocity and density profiles of particles across the flame propagation through a micro-iron dust cloud.

Calculations related to the two-phase flow considered in this study were performed with ANSYS FLUENT™ 13.0, which can associate a discrete phase and a structured turbulence model.

Reynolds Averaged Navier–Stokes (RANS) Model for Fluid Flow

Reynolds Stress Model (RSM) is a second order model for closure of the mathematical description of the fluid flow that averages the Navier–Stokes equations in a more structured and elaborated way than other RANS models. It is due to the inclusion of seven equations for description of turbulence in a 3D simulation whereas other models consider only 2, 3 or 4 equations for characterization of turbulence such as k–ε, k–kl–ω and Transition SST. The additional equations solve directly the transport equations. Kang and Choi (2004) developed comparative analyses of turbulence models in open-channel flows and identified better predictions of mean flow and turbulence statistics with RSM rather than with k–ε model, which overpredict the streamwise mean velocity.

On the one hand, the simple URANS methods are not appropriate for description of flows associated to complex stress fields with high magnitude because they are based on the assumption of isotropy on momentum transfer attributed to turbulent viscosity. On the other hand, the RSM allows considering directional effects in the stress fields by suppressing this assumption on the development of vortex in the fluid flow. This fact improves accuracy in the prediction of effects associated to streamlines curvature, flow rotation, shear rates and buoyancy effects due to the more rigorous method considered for turbulence.

The implementation of RSM constitutes an important factor for a better representation of the transient dispersion process. However, it is necessary to admit that the main disadvantage of this model lies on the higher computational cost in comparison with other URANS turbulence models discussed in this paper.

A case study has been defined with a simulation of solid rigid particles dispersion in a typical characterization test. The simulation considers 60 ms of dispersion process of micrometric glass beads. The flow domain, which is shown in Fig. 2, consists of a hemispherical base designed for initial location of dust deposits and a parallelepiped tube defined according to the geometrical specifications of the characterization apparatus.

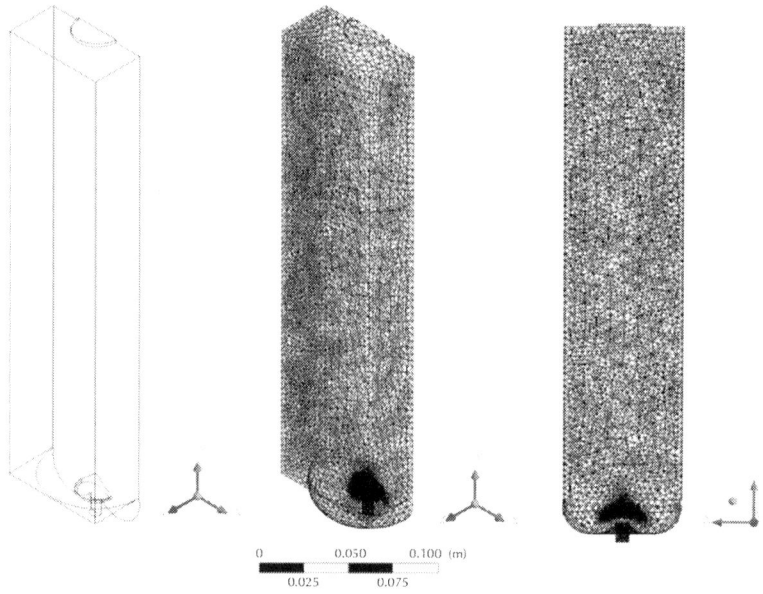

Figure 2: Description of the flow domain. (a) Gas injection, (b) Dispersion nozzle, (c) Outlet.

An injection of pressurized gas is performed through a nozzle located at the bottom of the tube, which also defines a flow restriction and determines the distribution of gas flow. For this reason, the mesh refinements in the regions were located near the tube walls and the nozzle. The mesh is composed of 245 572 nodes that constitute 499 677 elements. They have been defined according to a patch conforming meshing algorithm, implemented with tetrahedrons located on an unstructured grid. The size of the elements considered for the numerical solution of Navier–Stokes equations with the finite-volume method ranges between 0.17 mm near the dispersion nozzle and 3.32 cm in regions located the top of the tube.

After adjustment for the number of iterations at every time step, in order to improve the convergence level, an assessment of the required time for simulation process has been performed with three different turbulence models by considering the same computational resources. The preliminary analyses were performed with a server Intel Xeon X5650 with 2 processors of 2.66 GHz installed with a set of 48 GB of RAM. Fig. 3 shows that a two equations turbulence model (k–ε) needs about 0.042 h (2.52 min) per time step in an Eulerian–Lagrangian approach. Whereas a three equations model requires 0.061 h (3.65 min) and the seven equations model (RSM) needs 0.082 h (4.92 min) per time step. It represents an increase of 96% and 35.5% in the calculation time respectively to reach convergence levels below the tolerance limits (10^{-2}) for the flow variables analyzed.

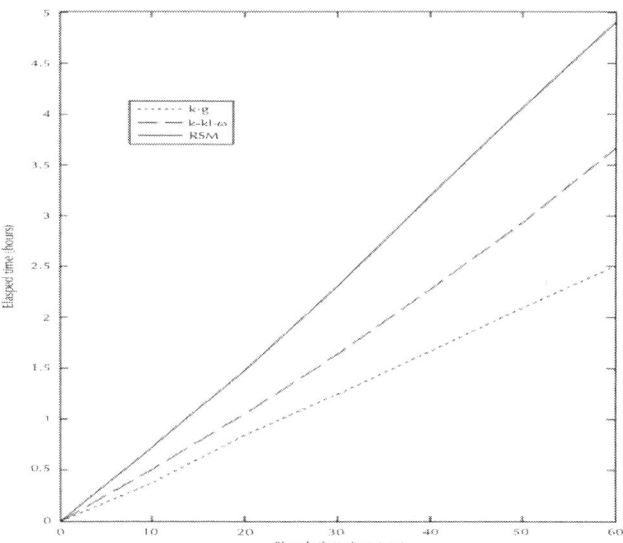

Figure 3: Comparison of required simulation time for analysis of dispersion of rigid particles as a function of the turbulence model.

Several factors have determined the convergence level and the computational cost of dispersion simulations. The process concerned in this paper possesses several specificities that imply the necessity of considering complex calculation methods. On the one hand, the analysis of a transient process with a dynamic boundary condition (gas injection) requires fine mesh grids for the calculation of fluid flow properties. This fact is due to constant changes in fluid flow momentum caused by the gas expansion and its subsequent reduction in velocity for a specific location inside the dispersion tube. For this reason, convergence and results accuracy have been enhanced by using the MUSCL method, which is a third order discretization alternative, for determining flow variables and solving turbulence equations. The scheme adopted by this alternative consists on an approximation of the flux crossing the interface of every finite-volume by employing two reconstructed values situated on both sides of the edge combined with a monotone numerical flux function. The values are recalculated according to a first computation of the flux through the meshing components and a subsequent modification of the gradients to consider a Total Variation Diminishing (TVD). The MUSCL technique is widely used in the industrial context due to its simplicity and adaptation capacity to respond to modeling evolutions and complexities (Buffard and Clain, 2010).

Moreover, the multi grid 'W-cycle' calculation scheme has been implemented. Previously, Rao and Medina (2003) have discussed the main characteristics of multigrid formulations in solutions of hyperbolic equations at a constant time step over a series of spatial domains. Analyses have been focused on the description of numerical results of computational studies that consider variations in grid spacing. This fact implies that the domain with the least grid spacing is addressed as fine domain and the coarse domains correspond to larger spacing values. The conclusions of this study establish that the computational effort required to solve the equations is reduced due to a lower number of grid nodes.

This process also involves a two differential equations system defined for every particle according to Newton's second law. Additionally, it is characterized by interaction between the two phases. This fact has been considered with the adjustment of time discretization with specific time steps for each phase. A time step of 1 ms has been defined for fluid flow while discrete phase equations have been integrated with a time step of 0.05 ms. Time integration parameters constitute an aspect of main interest for cases in which the length scales are very similar or when the fragmentation phenomenon is significant.

Reynolds Stress Model (RSM) for Description of Turbulence

After visualizing the velocity components in the flow field with the Reynolds decomposition $(u_i(x_i,t)) = \bar{u}_i(x_i) + u'(x_i,t))$, it is possible to determine the value of the kinetic turbulent energy as the trace of the stress tensor formed after averaging the Navier–Stokes equations.

$$k = \frac{1}{2}(u'^2_i + u'^2_j + u'^2_k) \quad (1)$$

The generalized representation of the Reynolds stress equations $(\rho_{u'_i u'_j})$ is defined by different terms associated to momentum transfer during the dispersion process.

$$\frac{\partial}{\partial t}\rho\overline{u'_i u'_j} + C_{ij} = D_{T,ij} + D_{L,ij} + P_{ij} + G_{ij} + \phi_{ij} + \varepsilon_{ij} + F_{ij} \quad (2)$$

Table 2 describes the parameters considered in Eq. (2):

Table 2: Parameters considered for the transport of Reynolds stresses with the RSM turbulence model

Non-modeled parameters in numerical solution	Modeled parameters in numerical solution
Convection $C_{ij} = \frac{\partial}{\partial x_k}(\rho u_k \overline{u'_i u'_j}) = \nabla \times (\rho \overline{u'_i u'_j} u)$ (3)	Turbulent diffusion. $D_{T,ij} = -\frac{\partial}{\partial x_k}[\rho \overline{u'_i u'_j u'_k} + \overline{p(\delta_{kj} u'_i + \delta_{ik} u'_j)}]$ (4)
Molecular diffusion $D_{L,ij} = \frac{\partial}{\partial x_k}\left[\mu \frac{\partial}{\partial x_k}(\overline{u'_i u'_j})\right]$ (5)	Buoyancy effects. $G_{ij} = -\frac{\mu_t}{\rho Pr_t}\left(g_i\frac{\partial \rho}{\partial x_j} + g_j\frac{\partial \rho}{\partial x_i}\right)$ (6)
Stress production $P_{ij} = -\rho\left(\overline{u'_i u'_k}\frac{\partial u_j}{\partial x_k} + \overline{u'_j u'_k}\frac{\partial u_i}{\partial x_k}\right)$ (7)	Pressure–strain. $\phi_{ij} = p\left(\frac{\partial u'_i}{\partial x_j} + \frac{\partial u'_j}{\partial x_i}\right)$ (8)
System rotation $F_{ij} = -2\rho\Omega_k(\overline{u'_j u'_m}\varepsilon_{ikm} + \overline{u'_i u'_m}\varepsilon_{jkm})$ (9)	Dissipation $\varepsilon_{ij} = -2\mu\overline{\frac{\partial u'_i}{\partial x_k}\frac{\partial u'_j}{\partial x_k}}$ (10)

In Eq. (6), μ_t and Pr_t are respectively the turbulent viscosity and the turbulent Prandtl number. These variables determine the buoyancy effects according to local variations in fluid density and the components of the gravitational acceleration (gi). Eq. (9) considers the rotation vector (Ω_k) and ε_{ijk}, which is a Levi–Civita factor in m direction that modifies its value according to the directions of velocity perturbations; ε_{ikm} is 1 if i, j and k are different and possess a cyclical order (even permutation), it will have a value of −1 if i, j and k are different and possess an anti-cyclical order (odd permutation) and will be 0 if two indexes are similar.

Lagrangian Approach for Dust Particles Tracking

The particles trajectories have been determined according to a force balance established on every particle that takes into account the different forces exerted on the particle during the dust dispersion.

$$m_p \frac{du_p}{dt} = m_p \left[F_D(u-u_p) + \frac{g_x(\rho_p - \rho)}{\rho_p} \right] \quad (11)$$

The drag force $m_p F_D(u-u_p)$ is calculated with Eqs. (12), (13) and (14) and the drag coefficient (C_D) is calculated according to the particle Reynolds number (Morsi and Alexander, 1972) with the parameters summarized in Table 3:

$$F_D = \frac{18\mu}{\rho_p d_p^2} \frac{C_D Re}{24} \quad (12)$$

$$Re_p = \frac{\rho d_p |u_p - u|}{\mu} \quad (13)$$

$$C_D = a_1 + \frac{a_2}{Re_p} + \frac{a_3}{Re_p^2} \quad (14)$$

Table 3: Drag coefficients defined according to the Morsi and Alexander's model

Range of particle Reynolds number	α_1	α_2	α_3
0.0–0.1	0.000	24	0
0.1–1.0	3.690	22.73	0.0903
1.0–10.0	1.2220	29.1667	−3.8889
10.0–100.0	0.6167	46.50	−116.67
100.0–1000.0	0.3644	98.33	−2 778
1000.0–5000.0	0.3570	148.62	−47 500
5000.0–10000.0	0.4600	−490.546	578 700
≥10000.0	0.5191	1 662.5	5 416 700

In Eqs. (12) and (13), ρ_p is the particle density and d_p represents its diameter. It should be highlighted that many additional forces, such as the effect of pressure gradients and fluid acceleration, can take part in the Newton's second law. Nevertheless, the micrometric size scale of the disperse phase poses a major influence of the particle density, which is much higher than the gas density (Feuillebois, 1980). As a consequence, only buoyancy and drag forces have been taken into account. However, due to the difference between particle density and gas density, buoyancy could have been neglected.

A two-way coupling is considered in the scheme to solve the discrete and continuous phase equations alternately until the solutions in both phases reach a convergence level. Additionally, the turbulent dispersion of solid particles is predicted with a stochastic tracking approach that integrates the trajectory equations according to the Reynolds decomposition along a particle path (s) during integration. In this order, the random effects of turbulence are established with a number representative of particles.

The numerical prediction of particles dispersion considers an integral time scale, which describes the time spent in turbulent motion along the particle path (s):

$$T = \int_0^\infty \frac{u'_p(t) u'_p(t+s)}{u'_p(t)^2} ds \qquad (15)$$

The particle dispersion rate is associated to this variable due to its representation of the turbulent motion in the flow. Moreover, the consideration of a movement along with the fluid of the smallest particles implies that their integral time becomes the fluid's Lagrangian integral time (TL):

$$T_L = C_L \frac{k}{\varepsilon} \qquad (16)$$

The value of the Lagrangian integral time can be determined for the RSM by matching the diffusivity of tracer particles to the scalar diffusion rate predicted by the turbulence model:

$$T_L \approx 0.3 \frac{k}{\varepsilon} \qquad (17)$$

After establishing a characteristic time scale, it is possible to pose a discrete random walk (DRW) model that simulates the interaction of a particle with a succession of discrete stylized fluid phase turbulent eddies, which are defined by the following conditions: (i) Gaussian distributed random velocity fluctuations u'_i, u'_j, u'_k, (ii) time scale (τ_e).

The velocity fluctuations that prevail during the lifetime of the turbulent eddy are sampled with a Gaussian probability distribution defined with a normally distributed random number (ζ):

$$u'_i = \zeta \sqrt{\overline{u'^2_i}} \qquad (18)$$

$$u'_j = \zeta \sqrt{\overline{u'^2_j}} \qquad (19)$$

$$u'_k = \zeta \sqrt{\overline{u'^2_k}} \qquad (20)$$

Particles Fragmentation: break-up Model

The breakup model considered for variations in particles size distribution defines a proportionality relationship between the diameter of the new formed particles and the wavelength of the fastest-growing unstable surface wave on the parent particle (Λ).

$$d_p = 2B_0 \Lambda \qquad (21)$$

In this order, the change rate of the radius in the parent particle (a) is given by the following expression, which is defined by the breakup time (τ) and the most unstable wave (Ω):

$$\frac{da}{dt} = -\frac{(a - d_p/2)}{\tau}, \quad \frac{d_p}{2} \leq a \qquad (22)$$

$$\tau = \frac{3.726 B_1 a}{\Lambda \Omega} \qquad (23)$$

where the variables Λ and Ω are defined by the following Eqs. (24) and (25):

$$\frac{\Lambda}{a} = 9.02 \frac{(1 + 0.450 h^{0.5})(1 + 0.4 Ta^{0.7})}{(1 + 0.87 We^{1.67})^{0.6}} \qquad (24)$$

$$\Omega \left(\frac{\rho_p a^3}{\sigma} \right) = \frac{(0.34 + 0.38 We^{1.5})}{(1 + Oh)(1 + 1.4 Ta^{0.6})} \qquad (25)$$

where Oh and Ta are the Ohnesorge and Taylor numbers, which are established by the Weber number of every phase and the particle Reynolds number (Re_p). These dimensionless numbers can be defined as follows:

$$Oh = \frac{\sqrt{\rho_p u^2 a / \sigma}}{Re_p} \qquad (26)$$

$$Ta = Oh\sqrt{\frac{\rho u^2 a}{\sigma}} = Oh \times We \tag{27}$$

In this model, mass is accumulated from the parent particle at a rate τ until the shed mass is equal to 5% of the initial particle mass. At this time, a new particle is created with a diameter *dp*. Only velocity and diameter change for the new particle. It is given a velocity component randomly selected in the plane orthogonal to the direction vector of the parent particle. The momentum of the parent particle is adjusted so that momentum is conserved.

EXPERIMENTAL RESULTS

An aspect of main interest discussed in this paper lies on the verification of the degree of homogeneity in the mixture by the description of segregation regions of the solid phase. For this reason, the discussion is focused on the description of gas injection and dispersion time, which must guarantee the necessary mixing level for ensuring the accuracy of safety parameters and tests reproducibility. Ignition generally occurring before the sedimentation phase, it has not been studied in detail.

Validation of the Test Procedure: Experiments with Glass Beads

The first experiments have been carried out with glass beads whose properties are summed up in Table 1. On Fig. 4, the evolution of the optical concentration at electrodes height during the dust dispersion is described.

Dust Explosions: CFD Modeling as a Tool to Characterize the ...   275

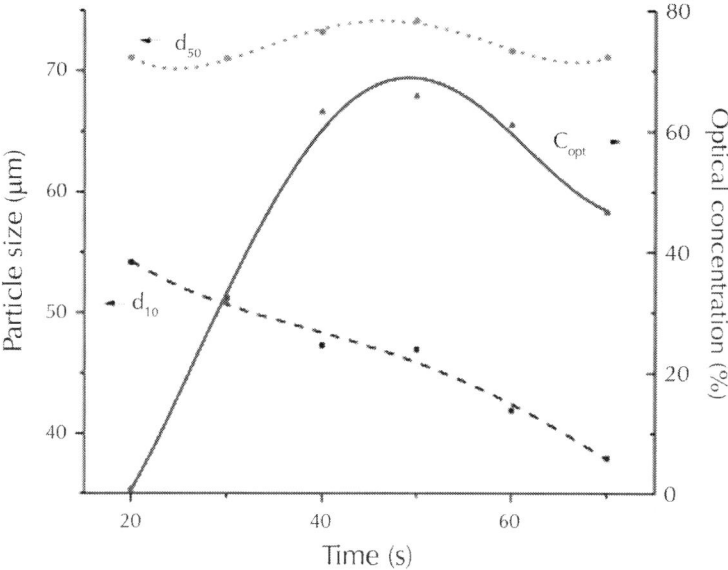

Figure 4: Evolution of characteristic diameters and of the optical concentration of the dust cloud during the dispersion of glass beads, at electrodes height −12.5 cm.

The variations observed in the optical concentration and the diameters d_{10} and d_{50} determine the behavior of the bulk of the dust cloud during the dispersion process. The profiles associated to the light scattering induced by the solids establish that some of the largest aggregates are affected by inertial effects, which increases their rising velocity. Afterwards, the other particles in the dust cloud reach the laser elevation and increase the optical concentration measured by the laser diffraction sensor. Moreover, Fig. 4 shows that the small particles are fluidized with a higher velocity, which implies a significant diminution in d_{10} during the dispersion process. However, it has a low impact on the global particle size distribution and the overall variations of the d_{50} are limited to 5 μm.

Finally, the dust cloud reaches higher elevations and the optical concentration reduces down to levels determined by the sedimentation velocities of the aggregates, which have been dispersed.

Aluminum Dusts Dispersions

Weiler et al. (2010) classified the hydrodynamic stresses exerted on the disperse phase surface according to the turbulence phenomenon by establishing the effects attributed to the gradients of velocity and collisions on solids fragmentation. For this reason, it is necessary to determine the instantaneous particle size distributions to identify variations that could have an influence on the experimental flammability characterization results.

The laser diffraction sensor has been positioned at the same height of ignition sources (electrodes) in order to compare the mean size determined by PSD analyses and the CFD simulations. Experimental tests have been performed with microsized powders.

Fig. 5 describes the dust cloud rising in the tube through the evolution of the solid fraction at a specific height of the dispersion tube (electrodes level). Fig. 5(a) and (b) show two different rising profiles that can be defined by their PSD and their dust concentrations. Both samples undergo an important reduction on their mean sizes during their dispersions. This fact evidences inertial effects during the first stage of dispersion that contribute to the fluidization of larger aggregates, with subsequent smaller variations due to the dusts sedimentation (Li and Ahmadi, 1992) and also to the aggregates fragmentation process, which will be discussed below.

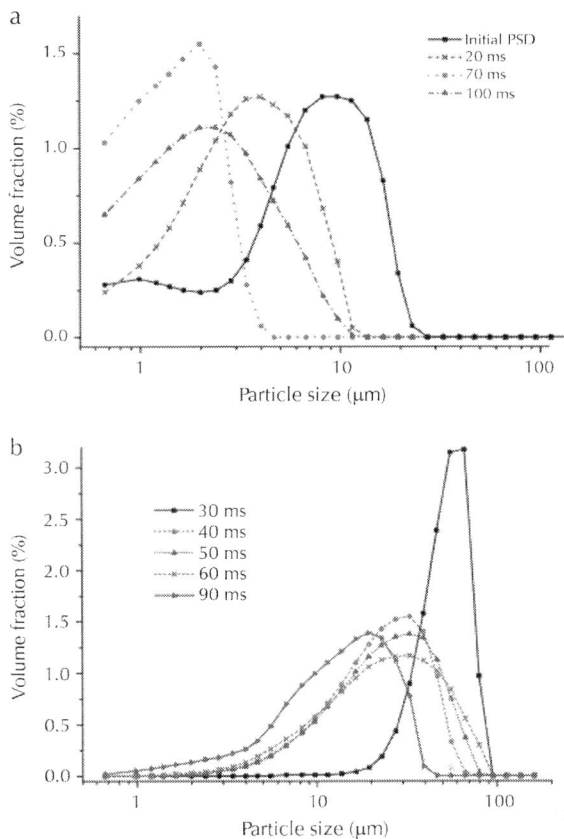

Figure 5: Evolution of the PSD of aluminum micrometric particles during their dispersion. (a) Aggregates with a PSD below 15 μm, (b) Aggregates with a PSD below 125 μm.

Aluminum Aggregates (Particles below 125 μm)

The combustible solid chosen for this study were micrometric aluminum particles. The analyzed samples are characterized by a PSD below 125 μm, with agglomerates diameters of approximately 80 μm. This material spontaneously forms aluminum oxide when exposed to the atmosphere. The formed film becomes an effective barrier to prevent further reaction of the aluminum with the environment. The induced ignition of this material is caused by a local disruption of the protective layer produced by the supply of sufficient energy (Chiffoleau et al., 2006).

Fig. 6 presents the initial stages of micrometric aluminum dispersion. Images have been taken with a high speed camera, which has been adjusted with a resolution of 1632×1200 pixels to register the two-phase flow development with 1016 frames per second. The internal conditions evidence a homogeneous distribution, in which the aggregates dispersed in the dust cloud rise with a characteristic profile that constitutes a flat front, which is significantly affected by the gas flow distribution after 40 ms of dispersion.

Figure 6: Micrometric aluminum dispersion inside the modified Hartmann tube. (a) 0 ms, (b) 10 ms, (c) 20 ms, (d) 30 ms, (e) 40 ms, (f) 50 ms, (g) 60 ms.

During the first 35 to 40 ms, the velocity of the dust cloud front is approximately constant. Then, the two-phase mixture is redistributed due to the vorticity induced by the walls and the high turbulence of the fluid. Weiler et al., (2010) establish that the PSD varies significantly during this process due to hydrodynamic stresses exerted by the gas flow. At the bottom of the tube, the injected air flows at transonic conditions and the solids particles possess low velocities that favor rotation and drag forces that reduce the mean size of the powder. Nevertheless, it is necessary to take into account that the collision probability at the bottom of the tube is higher. This fact constitutes an important factor for fragmentation and aggregation of solid phase.

DESCRIPTION OF DUSTS DISPERSION THROUGH THE CFD SIMULATION

The flow variables characterizing the two phases during the dispersion process have been described by the Euler–Lagrange approach implemented in the CFD simulation. These variables have been associated to the standardized operating procedures. Moreover, the trajectories followed by the disperse phase have been predicted in order to determine the segregation regions that constitute an important uncertainty factor of the flammability tests developed on combustible dusts.

Boundary Conditions of the Flow Domain

The internal and external faces of the flow domain have been classified according to the parameters set for the fluid flow development. The five regions shown in Fig. 7 represent the conditions which applied to the continuous phase. It is necessary to underline that these conditions are time-dependant because the gas conditions vary during the injection process.

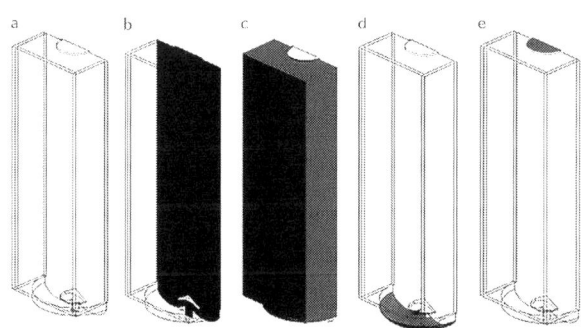

Figure 7: Boundary conditions of the flow domain.

Zones (a) and (e) represent respectively the gas inlet and outlet. The outlet boundary condition (e) has been expressed by considering that the pressure outside the tube was constant and equal to the atmospheric pressure. The inlet boundary condition (a) has been established with a user defined function (UDF—Ansys Fluent) that characterizes the pressure evolution during the injection process:

$$p = P_{atm} + P_{injection} \times (e^{-t^2/2t\sigma^2}) \quad (28)$$

The standard deviation ( ) has been determined by considering that the injection pressure $P_{injection}$ is reduced at 0.5% of its initial value after the first 30 milliseconds of the injection process, which is experimentally proven.

Fig. 7(b) represents a symmetry condition that was considered after evidencing the development of gas flow streamlines shown in Fig. 8. Alternative simulations took into account the complete flow domain and posed symmetric profiles at regions near the ignition electrodes.

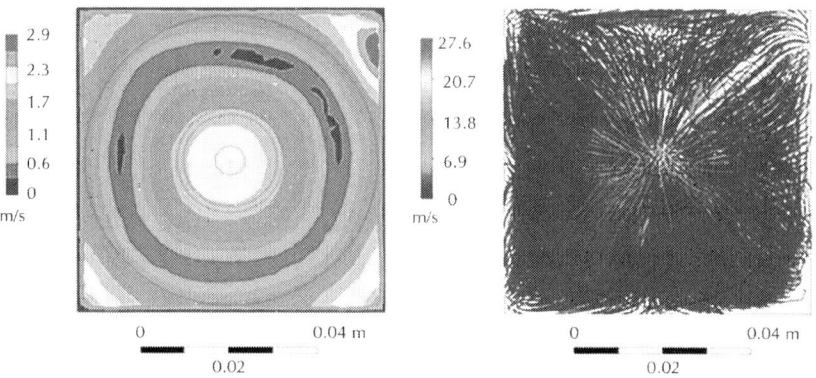

Figure 8: Velocity magnitude and gas flow streamlines at 5 cm after 60 ms of dispersion.

Finally, the tube walls shown on Fig. 7(c) and (d) have been defined with the no-slip condition for the gas flow and a reflection condition for the solid particles. Besides, the bottom of the tube (c) is used for the initial localization of the discrete phase.

Initial Conditions of the Flow Domain

The initial conditions of the flow domain correspond to the laboratory environment: the magnitude of the velocity field is considered to be null and the total pressure is the atmospheric pressure. Additionally, the discrete phase model, which predicts the trajectories of the dust particles, is characterized by a gas injection that is distributed along the surface shown in Fig. 7(d). 0.80 g of powder are initially deposited on this surface.

Transient Dispersion Process

The transient process considered in this study is affected by several characteristics of the flow domain and the boundary conditions defined for gas injection. For this reason, an analysis of continuous phase development has been defined in order to establish the main conditions of the fluid flow inside the tube. The injection nozzle located at the bottom of the tube constitutes the main factor for gas and solids distribution in the confined flow at the lower locations inside the tube. During the dispersion process, important variations in gas velocity profiles have been evidenced at various locations. This behavior was not similar to a typical gas internal flow. After considering the end of gas injection (30 ms), the velocity magnitudes inside the recipient were determined. The behavior shown in Fig. 9a and was termed as "gulf-effect" (Bahramian and Mansour, 2010). These authors established that two-phase internal flows developed in injections tend to distribute near the tube walls and later favor the location of solid particles in the middle of the equipment. Additionally, Huilin et al. (2010) observed the development of core-annular flows in an analysis of solid dispersions in gases that can be explained by this phenomenon as well.

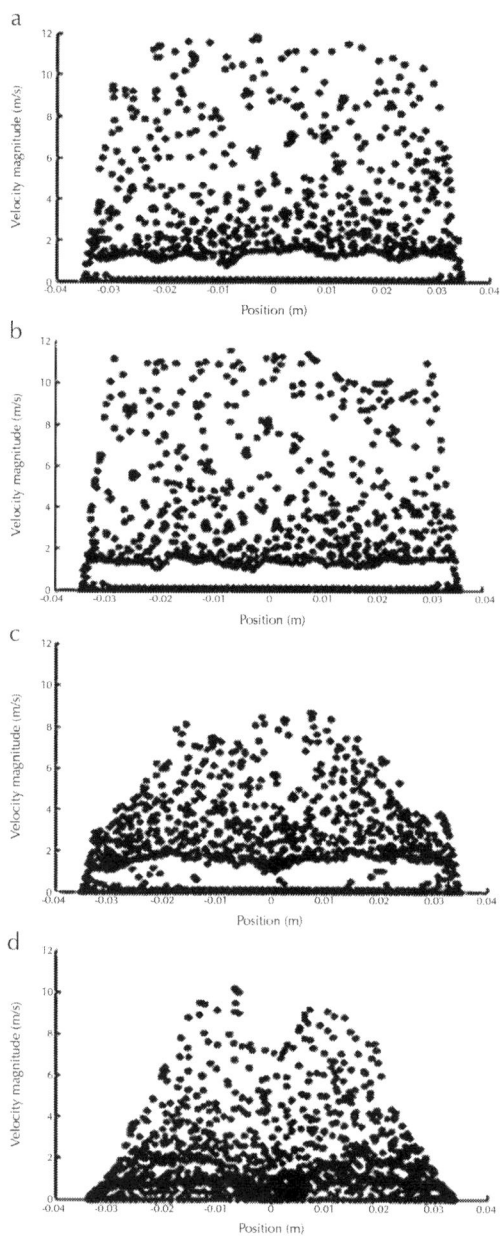

Figure 9: Velocity magnitudes at different heights of the tube at 30 ms. (a) 10 cm over nozzle (b) 5 cm over the nozzle (c) 2 cm over the nozzle (d) Nozzle elevation (0 cm).

Dust Explosions: CFD Modeling as a Tool to Characterize the ... 283

The upper regions inside the vessel evidence more uniform velocity profiles. This fact can be attributed to the variation in the transversal area of the dispersion tube and a decrease in pressure along the vessel caused by the gas expansion. After considering the development of the internal gas flow, it is possible to determine its influence on the segregation levels of the combustible dust analyzed during the test. Fig. 12 presents the concentration profiles at different planes or heights inside the dispersion tube. The variation in the transversal section defines a uniform concentration profile that implies an additional factor for description of the development of the dust cloud that was evidenced experimentally during the dispersion process.

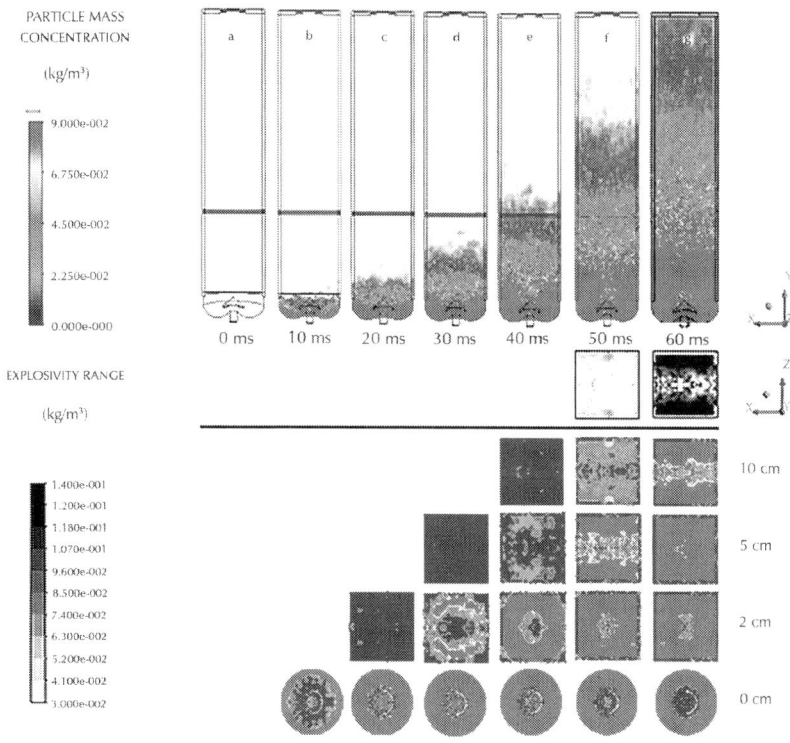

Figure 10: Concentration profiles at different heights inside the Hartmann tube during dispersion process. The reference (0 cm) is located at the top of the dispersion nozzle.

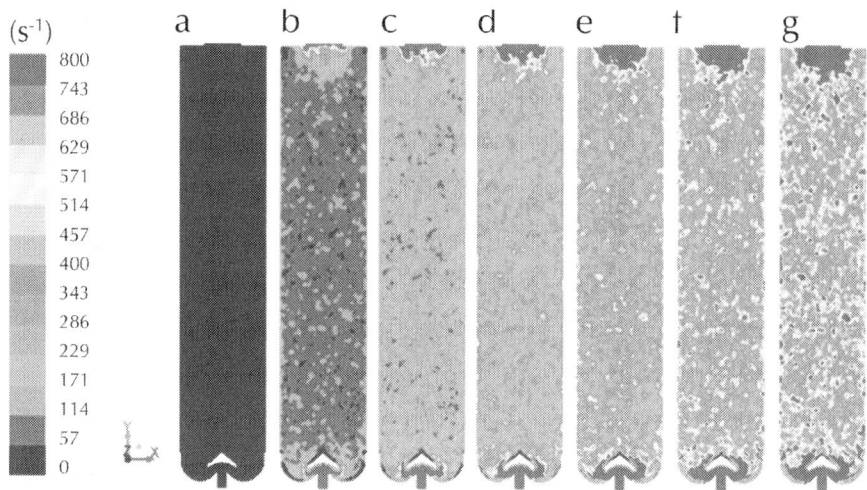

Figure 11: Vorticity magnitude of the gas flow during the dispersion process. (a) 0 ms, (b) 10 ms, (c) 20 ms, (d) 30 ms, (e) 40 ms, (f) 50 ms, (g) 60 ms.

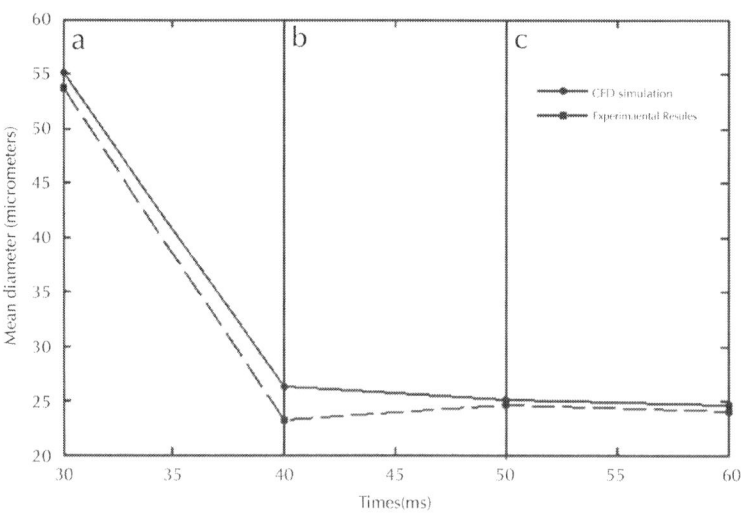

Figure 12: Mean diameter of 'Micro-Al 42' at ignition sources location during dispersion process. (a) Instability region, (b) Transition region, (c) Stability region.

The minimum explosive concentration of the micrometric aluminum dust discussed in this paper has been determined experimentally in the standardized 20 l sphere according to the operating specifications of the equipment. The experimental result determined for this aluminum dust sample was 0.09 kg/m^3. Nevertheless, different values that range between 0.03 and 0.14 kg/m^3 have been reported for various aluminum dusts (Dufaud et al., 2010). This study shows that, for this specific powder, MEC can be reached at the electrodes height approximately after 60 ms. It implies that the ignition delay *tv* should not be lower than this value for such powders.

The vorticity of the fluid is affected considerably by the shape and location of the dispersion nozzle due to the stresses induced by the internal walls of the tube. On the other hand, the consideration of the vorticity levels at different elevations inside the tube, the highest vorticity levels are near the walls. However, the magnitude differences among the different locations at a specific height are not significant. The profiles shown in Fig. 11 explain the behavior of the aluminum concentration during the period of time elapsed between 40 and 60 ms, in which, the segregation of solid phase reduces to constitute profiles that describe the behavior observed experimentally. The CFD simulation and the characterization performed with the high-speed camera pose an interesting hypothesis about the location of the ignition sources for the characterization of combustible solids, because an intermediate height of the electrodes guarantees uniformity in the distribution of discrete phase and a high concentration for measurement. Furthermore, it is necessary to take into account that it is not possible to identify regions with high vorticity levels during the first 30 ms seconds of dispersion.

Simulations also show that the area-weighted average turbulent kinetic energy raises up to 250 m^2/s^2 at approximately 35 ms and then decreases rapidly. This trend is highly correlated to the fragmentation process, which can be observed on Fig. 12. Moreover, the solid loading has a noticeable influence on the turbulence decay. An extension comparison of experimental data obtained by particle image velocimetry (PIV) and CFD simulation has been done for various powders (glass beads, aluminum, starch...). These results will be published in another paper.

Modeling of the Fragmentation Process

The fragmentation of solid particles inside the dispersion tube has been implemented by the adjustment of the parameter B_0 in Eq. (22). The PSD that was determined experimentally with the granulometric measurements was established in intervals of 10 ms since the arrival of the dust cloud to the elevation of the ignition sources. This variable modified its value from 13 at the beginning of granulometric analyses (25 ms) to 0.2 at the end of the dispersion process (60 ms). Fig. 12 presents the variations in the mean diameter of the dust cloud during the development of the two-phase flow. Initially, the sample has been characterized by determination of PSD in a sample in absence of forces but the gravitational. The values determined by the granulometric analysis have been summarized in Table 4 and have been adjusted with the Rosin–Rammler diameter distribution method, which is described with Equation 30, where, Y_d is the mass fraction of solids larger than the diameter d and the size constant $\bar{d}$ and the size distribution parameter (n) are 64.91 μm and 3.489, respectively.

$$Y_d = e^{-(d/\bar{d})^n} \tag{29}$$

Table 4: Al particle size distribution by granulometric analyses in absence of external stresses (micro-Al 42)

Mass fraction (%)	Diameter (μm)
0–10	26.63
0–16	33.18
0–50	57.41
0–84	76.75
0–90	80.60
0–99	86.36

The identification of initial conditions of the PSD was performed, and the evolution of mean size distribution at the ignition electrodes elevation was characterized. The profile, shown in Fig. 12, evidences a reduction of the analyzed variable of the tested sample during the first 60 ms of dispersion, because the initial mean diameter is 57.41 μm and the distribution has decreased to a mean value of 23 μm. The results obtained with the CFD simulation and the experimental analyses have allowed classifying the dispersion process in three different stages. Initially, the drag force exerted by the fluid flow on the surface of the particles causes a dramatic decrease in size distribution. This result can be explained after considering the drastic variation of internal gas conditions determined by the high pressure of the gas. The results define a period, in which, the eventual measurements of the explosivity parameters of the combustible material might be constrained to uncertainty factors attributable to the low vorticity levels at high elevations that imply a low collision probability of disperse particles. Afterwards, the diminution in the mean diameter of the PSD is restricted by the arrival of the homogeneous dust cloud and the development of the internal gas flow. Evidently, since the arrival of the bulk of the dust cloud and before sedimentation, the determination of explosivity parameters can be performed with a more stable profile.

According to the results obtained by Bouillard et al. (2010) and Washburn et al. (2010) the aluminum particles with diameters less than 20 μm have considerable differences in mechanisms of combustion because the combustion time of the material is determined by the kinetics of the chemical reaction rather than oxygen diffusion. Both particle size distributions and initial turbulence having great influences on the dust reactivity, the choice of a pertinent ignition delay is essential. For instance, in the case of the previously quoted aluminum powders, an analysis performed on variation of mean size diameter should recommend a period longer than 50 ms for the ignition delay. Thus, an analysis performed after this time can provide more precise and stable results for the minimum ignition energy and minimum explosive concentration (MIE and MEC).

Such analysis, combining both simulation and experiments, is recommended for every test of a combustible solid in order to consider the influence of its material properties (bulk density, rigidity, shape factor, etc.) on the dust cloud development, and therefore, on the flammability parameters. Besides, the analysis of various dispersion conditions (notably pressures) or dispersion nozzles can be considered as an aspect of interest for a more accurate description of internal two-phase flows in such apparatuses.

CONCLUSIONS

A flat profile has been evidenced for the flow behavior during the initial stages of aluminum dust dispersion inside the modified Hartmann tube. This fact has been explained by describing the development of the internal gas flow and the presence of the fragmentation mechanisms in the discrete phase. Additionally, the results pose the important characteristics of the influence of the variation of the geometry of the flow domain and the air injection on the distribution of solid phase and the development of internal gas flow. For aluminum powders, the evidences validate the previous assumption of mixture homogeneity for regions located at 10 cm over the nozzle, after identifying high segregation levels only at the bottom of the apparatus. For this reason, this elevation is recommended for the installation of ignition sources in order to reduce the uncertainty level associated to a combustible dust characterization test.

The results have also allowed identifying three different stages for dust dispersion according to the variation in the mean diameter of discrete phase at the ignition sources elevation. The results show a drastic variation of PSD and gas flow conditions during the first 40 ms of dispersion. However, the variation is not significant after this period of time.

According to the results obtained by simulation and experimental tests, a minimum ignition delay of 50 to 60 ms has been recommended after taking into account the variations in the vorticity of the fluid flow, the segregation levels of discrete phase and the mean diameter of the tested sample. However, it should be highlighted that these results are only valid for a specific powder and that changes in the material properties will lead to different dispersion characteristics. As a consequence, it seems that such integrated approach coupling simulation and experimental study, should be systematically applied to determine the most relevant (from a process point of view) or the most conservative conditions for dust flammability characterization.

REFERENCES

1. Amyotte, P.R., Pegg, M.J., Khan, F.I., 2009. Application of inherent safety principles to dust explosion prevention and mitigation. Process Safety and Environment Protection 87, 35–39.
2. Bahramian, Reza A., Mansour, K., 2010. CFD modeling of TiO2 nano-agglomerates hydrodynamics in a conical fluidized bed unit with experimental validation. Iranian Journal of Chemistry & Chemical Engineering 29 (2), 105–120.
3. Bidabadi, M., Haghiri, A., Rahbari, A., 2010. Mathematical modeling of velocity and number density profiles of particles across the flame propagation through a micro-iron dust cloud. Journal of Hazardous Materials 176 (1-3), 146–153.
4. Bouillard, J., Vignes, A., Dufaud, O., Perrin, L., Thomas, D., 2010. Ignition and explosion risks of nanopowders. Journal of Hazardous Materials, 873–880.
5. Buffard, T., Clain, S., 2010. Monoslope and multislope MUSCL methods for unstructured meshes. Journal of Computational Physics 229 (10), 3745–3776.
6. Callé, S., Klaba, L., Thomas, D., Perrin, L., Dufaud, O., 2005. Influence of the size distribution and concentration on wood dust explosion: experiments and reaction modelling. Powder Technology 157, 144–148.

7. Chiffoleau, G., Newton, B., Holroyd, N.J.H., Havercroft, S., 2006. Surface ignition of aluminum in oxygen. Journal of ASTM International 3, 5.
8. Di Benedetto, A., Russo, P., Sanchirico, R., Di Sarli, V., 2013. CFD simulations of turbulent fluid flow and dust dispersion in the 20 l explosion vessel. Wiley Online Library 59 (7), 2485–2496.
9. Di Sarli, V., Russo, P., Sanchirico, R., Di Benedetto, A., 2013. CFD simulations of the effect of dust diameter on the dispersion in the 20 l bomb. Chemical Engineering Transactions 31, 727–732.
10. Dufaud, O., Traoré, M., Perrin, L., Chazelet, S., Thomas, D., 2010. Experimental investigation and modelling of aluminum dusts explosions in the 20 l sphere. Journal of Loss Prevention in the Process Industries 23 (2), 226–236.
11. Eckhoff, R., 1996. Prevention and mitigation of dust explosions in the process industries: a survey of recent research and development. Journal of Loss Prevention in the Process Industries 9 (1), 3–20.
12. Eckhoff, R., 2009. Understanding dust explosions. The role of powder and science technology. Journal of Loss Prevention in the Process Industries, 105–116.
13. Eckhoff, R.K., 1977. Pressure development during explosions in clouds of dusts from grain, feedstuffs and other natural organic materials. Fire Safety Journal 1 (2), 71–85.
14. Eckhoff, R.K., 2003. Assessment of ignitability, explosibility, and related properties of dusts by laboratory-scale tests. Dust Explosions in the Process Industries (third ed., pp. 473–548). Gulf Professional Publishing.
15. Erlend, R., Eckhoff, R., 2006. Initiation of dust explosions by electric spark discharges triggered by the explosive dust cloud itself. Journal of Loss Prevention in the Process Industries 19 (2-3), 154–160.

16. Feuillebois, F., 1980. Certains Problèmes d'écoulements Mixtes Fluides-particules. Paris VI. Ph.D. Thesis. Université Pierre et Marie Curie.
17. Glarner, T., 1984. Mindestzündernergie-Einfluss der Temperatur. VDI-Berichte [VDI-Verlag GmbH Düsseldorf] 494, 109–118.
18. Huilin, L., Shuyan, W., Jianxiang, Z., Gidaspow, D., Ding, J., Xiang, L., 2010. Numerical simulation of flow behavior of agglomerates in gas–cohesive particles fluidized beds using agglomerates-based approach. Chemical Engineering Science 65 (4), 1462–1473.
19. IEC 1241-2-3 Ed. 1.0 b., 1994. Electrical apparatus for use in the presence of combustible dust – Part 2: Test methods – Section 3: Method for determining minimum ignition energy of dust/air mixtures. Geneva: International Electrotechnical Commission.
20. ISO 6184-1., 1985. Explosion Protection Systems—Part 1: Determination of explosion indices of combustible dusts in air. International Organization for Standardization.
21. Jenkins, C.M., Ripley, R.C., Wu, C.-Y., Horie, Y., Powers, K., Wilson, W.H., 2013. Explosively driven particle fields imaged using a high speed framing camera and particle image velocimetry. International Journal of Multiphase Flow 51, 73–86.
22. Kalejaiye, O., Amyotte, P.R., Pegg, M.J., Cashdollar, K.L., 2010. Effectiveness of dust dispersion in the 20-l Siwek chamber. Journal of Loss Prevention in the Process Industries 23 (1), 46–59.
23. Kang, H., Choi, S.-U., 2004. Reynolds stress modeling of vegetated open-channel flows. Journal of Hydraulic Research 42 (1), 3–11.
24. Kosinski, P., 2011. Explosion suppression by a cloud of particles: numerical analysis of the initial processes. Applied Mathematics and Computation 217 (11), 5087–5094.

25. Li, A., Ahmadi, G., 1992. Dispersion and deposition of spherical particles from point sources in a turbulent channel flow. Aerosol Science and Technology 16 (4), 209–216.
26. Matsuda, O.-S., Yashima, M., Matsuda, T., Matsui, H., Miyake, A., Ogawa, T., 2001. A study of flame propagation mechanisms in lycopodium dust clouds based on dust particles' behavior. Journal of Loss Prevention in the Process Industries 14 (3), 153–160.
27. Morsi, S., Alexander, A., 1972. An investigation of particle trajectories in two-phase flow systems. Journal of Fluid Mechanics 55 (2), 193–208.
28. Rao, P., Medina, M., 2003. Evaluation of V and W multiple grid cycles for modeling one and two-dimensional transient free surface flows. Applied Mathematics and Computation 138, 151–167.
29. Sanchirico, R., Di Benedetto, A., Garcia-Agreda, A., Russo, P., 2011. Study of the severity of hybrid mixture explosions and comparison to pure dust–air and vapour–air explosions. Journal of Loss Prevention in the Process Industries 24, 648–655.
30. Trunov, M.A.T., 2006. Effect of Polymorphic Phase Transformations within an Alumina Layer on the Ignition of Aluminum Particles. Ph.D. Thesis. Department of Mechanical Engineering, New Jersey Institute of Technology.
31. Jian, W., Xinguang, L., Shengjun, Z., Radandt, S., Fuli, W., Chunli, R., 2009. Flame propagation through potato starch/air mixture in pipe of interconnected vessels. In: Proceedings of the Eighth International Conference of Granular Materials, 376–380.
32. Washburn, E., Webb, J., Beckstead, J., 2010. The simulation of the combustion of micrometer-sized aluminum particles with oxygen and carbon dioxide. Combustion and Flame 157 (3), 540–545.

33. Weiler, C., Wolkenhauer, M., Trunk, M., Langguth, P., 2010. New model describing the total dispersion of dry powder agglomerates. Powder Technology 203 (2), 248–253.

Citations

CHAPTER 1

Rolf K. Eckhoff, "Dust Explosion Prevention and Mitigation, Status and Developments in Basic Knowledge and in Practical Application," International Journal of Chemical Engineering, vol. 2009, Article ID 569825, 12 pages, 2009. doi:10.1155/2009/569825.

CHAPTER 2

Joseph Kalman, Nick G. Glumac, and Herman Krier, "Experimental Study of Constant Volume Sulfur Dust Explosions," Journal of

Combustion, vol. 2015, Article ID 817259, 11 pages, 2015. doi:10.1155/2015/817259.

CHAPTER 3

Junaid Hassan, Faisal Khan, Paul Amyotte, Refaul Ferdous, Industry specific dust explosion likelihood assessment model with case studies, Journal of Chemical Health and Safety, Volume 21, Issue 2, March–April 2014, Pages 13-27, ISSN 1871-5532, http://dx.doi.org/10.1016/j.jchas.2013.11.005.

CHAPTER 4

Niansheng Kuai, Jianming Li, Zhi Chen, Weixing Huang, Jingjie Yuan, Wenqing Xu, Experiment-based investigations of magnesium dust explosion characteristics, Journal of Loss Prevention in the Process Industries, Volume 24, Issue 4, July 2011, Pages 302-313, ISSN 0950-4230, http://dx.doi.org/10.1016/j.jlp.2011.01.006.

CHAPTER 5

A. Di Benedetto, P. Russo, P. Amyotte, N. Marchand, Modelling the effect of particle size on dust explosions, Chemical Engineering Science, Volume 65, Issue 2, 16 January 2010, Pages 772-779, ISSN 0009-2509, http://dx.doi.org/10.1016/j.ces.2009.09.029.

CHAPTER 6

Vimlesh Kumar Bind, Shantanu Roy, Chitra Rajagopal, A reaction engineering approach to modeling dust explosions, Chemical Engineering Journal, Volumes 207–208, 1 October 2012, Pages 625-634, ISSN 1385-8947, http://dx.doi.org/10.1016/j.cej.2012.07.026.

CHAPTER 7

M. Silvestrini, B. Genova, F.J. Leon Trujillo, Correlations for flame speed and explosion overpressure of dust clouds inside industrial enclosures, Journal of Loss Prevention in the Process Industries, Volume 21, Issue 4, July 2008, Pages 374-392, ISSN 0950-4230, http://dx.doi.org/10.1016/j.jlp.2008.01.004.

CHAPTER 8

Carlos Murillo, Olivier Dufaud, Nathalie Bardin-Monnier, Omar López, Felipe Munoz, Laurent Perrin, Dust explosions: CFD modeling as a tool to characterize the relevant parameters of the dust dispersion, Chemical Engineering Science, Volume 104, 18 December 2013, Pages 103-116, ISSN 0009-2509, http://dx.doi.org/10.1016/j.ces.2013.07.029.

Index

A

Activation energy 176, 182, 183, 188, 190
Aluminum combustion 173

B

Bulk density 177, 178

C

Coefficient of correlation (CoC) 183
Coefficient of correlations (CoCs) 189

Computational fluid dynamics (CFD) , 20, 7
Computational Fluid Dynamics (CFD) 260
Correlations 202, 208, 210, 229, 247

D

Direct proportion 135
Discrete random walk (DRW) 272
Dust Explosion Simulation Code (DESC) 7
Dust flame 39

E

Experimental data 168, 183, 184, 185, 186, 188, 189, 190, 193, 194, 195, 196, 197
Experimental study 195
Extensive work 169

F

FLame ACceleration Simulator (FLACS) 7

I

Ignition energy 107, 108, 109, 110, 111, 118, 120, 122, 123, 124, 125, 126, 127, 128, 129, 134, 135
Industrial enclosures 231, 246, 297
Initial temperature 117, 123, 124, 128, 134, 135
International Electrotechnical Commission (IEC) 16

L

Lower explosion limit (LEL) 108
Lower explosive limit (LEL) 12

M

Minimum Explosive Concentration (MEC) 259
Minimum ignition energies (MIEs) 13
Minimum Ignition Energy (MIE) 258, 259

N

National Fire Protection Association (NFPA) 73
Natural Sciences and Engineering Research Council (NSERC) 104

O

Omega data acquasition (DAQ) 46
Overpressure 202, 212, 215, 223, 231, 232, 233, 234, 235, 239, 240, 244, 245, 246, 247, 297
Oxygen concentration 129, 130

P

Particle size distribution (PSD) 257
Perfectly-stirred reactor (PSR) 61
Photomultiplier tubes (PMT) 46

R

Real industrial plant 2
Relevant physical 3
Reynolds Stress Model (RSM) 265, 269

S

Scanning electron microscope (SEM) 42
Sulfur dust 40
Surface reaction 169, 174, 175, 176, 188

T

Thermal radiation 5, 14
Total Variation Diminishing (TVD) 268

V

Volume fraction 172, 173, 174, 177, 178
Volumetric ignition phase 123